2nd International Conference on Architecture and Civil Engineering (ICACE 2018)

AA006052

IOP Conference Series: Materials Science and Engineering
Volume 401

Batu Ferringhi, Malaysia
9 - 10 May 2018

Editors:

Nordin Abd Rahman
Sattar S. Emamian
Mokhtar Awang

ISBN: 978-1-5108-7255-4
ISSN: 1757-8981

Printed from e-media with permission by:

Curran Associates, Inc.
57 Morehouse Lane
Red Hook, NY 12571

Some format issues inherent in the e-media version may also appear in this print version.

Copyright© (2018) by the Institute of Physics
All rights reserved. The material featured in this book is subject to IOP copyright protection, unless otherwise indicated.

Printed by Curran Associates, Inc. (2018)

For permission requests, please contact the Institute of Physics at the address below.

Institute of Physics
Dirac House, Temple Back
Bristol BS1 6BE UK

Phone: 44 1 17 929 7481
Fax: 44 1 17 920 0979

techtracking@iop.org

Additional copies of this publication are available from:

Curran Associates, Inc.
57 Morehouse Lane
Red Hook, NY 12571 USA
Phone: 845-758-0400
Fax: 845-758-2633
Email: curran@proceedings.com
Web: www.proceedings.com

Table of contents

Volume 401

2nd International Conference on Architecture and Civil Engineering (ICACE 2018)

9–10 May 2018, Batu Ferringhi, Penang, Malaysia

Accepted papers received: 27 July 2018
Published online: 2 October 2018

Preface

Papers

IOP Conf. Series: Materials Science and Engineering **401** (2018) 011001 doi:10.1088/1757-899X/401/1/011001

Preface:

This volume presents a compilation of research works covering the fields of architecture and civil engineering. All manuscripts in this volume were presented during the 2nd International Conference on Architecture and Civil Engineering (ICACE 2018) which was held at Parkroyal Hotel, Penang, Malaysia on 09-10 May 2018. The editor(s) of the proceeding would like to express the utmost gratitude and thanks to all reviewers in the technical team for making this volume a success.

Editors:

Dr. Nordin Abd Rahman

Dr. Sattar S. Emamian

Ir. Dr. Mokhtar Awang, CEng

Content from this work may be used under the terms of the Creative Commons Attribution 3.0 licence. Any further distribution of this work must maintain attribution to the author(s) and the title of the work, journal citation and DOI.
Published under licence by IOP Publishing Ltd

Content from this work may be used under the terms of the Creative Commons Attribution 3.0 licence. Any further distribution of this work must maintain attribution to the author(s) and the title of the work, journal citation and DOI.
Published under licence by IOP Publishing Ltd

IOP Conf. Series: Materials Science and Engineering **401** (2018) 011003 doi:10.1088/1757-899X/401/1/011003

Peer review statement

All papers published in this volume of *IOP Conference Series: Materials Science and Engineering* have been peer reviewed through processes administered by the proceedings Editors. Reviews were conducted by expert referees to the professional and scientific standards expected of a proceedings journal published by IOP Publishing.

Content from this work may be used under the terms of the Creative Commons Attribution 3.0 licence. Any further distribution of this work must maintain attribution to the author(s) and the title of the work, journal citation and DOI.
Published under licence by IOP Publishing Ltd

IOP Conf. Series: Materials Science and Engineering **401** (2018) 012001 doi:10.1088/1757-899X/401/1/012001

The Tectonic Expression: 'Reading' the works of Wooi Lok Kuang

V Ng[1] and A Wong[2]

[1] Associate Professor, School of Architecture, Building and Design, Taylor's University, Malaysia
[2] Postgraduate, School of Architecture, Building and Design, Taylor's University, Malaysia

E-mail: ng.foongpeng@taylors.edu.my

Abstract. In an age of commercialism and consumerism, there has been a notable demise in the employment of a tectonic expression in the making of architecture. This is due in part to a culture of mass production to meet the rising needs of consumers, resulting in a shift in the emphasis of the value of detailing and construction as an architectural expression. Instead, there arose a preoccupation with pastiche ornamentation, reducing the significance and meaning within detailing. Furthermore, construction began to be viewed as a means to achieve spatial effect as an end. However, the need for a tectonic expression in architecture has been highlighted by Western key thinkers such as Kenneth Frampton, Vitorrio Gregotti and Marco Frascari, who have stated that tectonic can be employed as a resolution to the degeneration of culture as a result of homogeneity of architecture. This is because tectonic involves the poetic amplification of construction knowledge which stems from culture and craftsmanship, and an appropriate application of tectonic in construction can lead to the construal of meaningful architecture. In Asia (Malaysia), there is limited study on the meaning of craft that stems from the richness of local cultures in relation to tectonics. By adopting a mixed-methodology of case study and interview, this paper studies works of Wooi Lok Kuang, a seminal Malaysian architect published for his folio of timber architecture, namely, Wooi Residence (2004), Ting Residence (2010) and House 9 (2015). It situates Wooi's design methodology of assembly of 'joints' in an Asian/Malaysian context within the larger body of knowledge regarding tectonics as found in the western literature and argues that Wooi's design methodology that is rich in cultural and contextual influences expounds on Frascari's framework of 'joints', including key factors that are specific to Wooi's architecture and the Malaysian context.

1. Introduction

In order to meet the demands of growing consumerism, the built environment has progressed towards industrial building technology and mass production, rendering the once eloquent craftsmanship of architecture to homogeneity. The current trend in making buildings amortizable has been achieved through the subconscious reduction of architecture to scenography (Frampton, 1990). The art of making buildings has lost its expressive design content, resulting in the lack of articulation of construction. Tectonics, which is the poetic application of construction, involves a consciousness of meaning within the applied construction technique. This meaning stems from the unique way in which it has been worked

Content from this work may be used under the terms of the Creative Commons Attribution 3.0 licence. Any further distribution of this work must maintain attribution to the author(s) and the title of the work, journal citation and DOI.
Published under licence by IOP Publishing Ltd

IOP Conf. Series: Materials Science and Engineering **401** (2018) 012001 doi:10.1088/1757-899X/401/1/012001

within a specific culture and context. The return to craftsmanship requires a reemphasis of the application of tectonics as a result of reconciliation between architecture and culture as expressed through construction details in joints. Architecture should be concerned with not only the pragmatic sense of design, but also with the symbolic significance of being situated in a culture-rich context.

A contemporary reflection on architecture posits tectonics, as an anachronistic notion (Garland, 2003). With a growing consumer culture, architecture developed towards industrial building where tectonic expression is replaced with modern technology as a consequence of mass production (Nilsson, 2007). Economic considerations became the emphasis of architecture in order to capitalize on the process of mass production, giving rise to homogeneity and devaluing the poetic articulation of tectonics. Furthermore, the success of Robert Venturi's decorated shed (Venturi, Brown, & Izenour, 1977), which portrays an application of independent ornamentation to a system of space and structure, prioritizes spatial expression as an end of the production of architecture. Kenneth Frampton's seminal work *Studies in tectonic culture: the poetics of construction in nineteenth and twentieth century architecture* (1990) summarises this decline in a tectonic expression as being the result of two correlating factors; a) mass production of architecture to meet the demands of consumerism, and b) spatial tendencies as an end in architecture.

With the rise of mass production, the development of technology has streamlined the production of architecture. Gregotti (1996) notes a loss of actual practice, tradition and knowledge of craftsmanship within architecture with the advent of a digital culture. It was not a matter of the elimination of details, but rather the 'alteration of the hierarchy of detail to whole', with less emphasis on the value of construction as an architectural expression. Marco Frascari (1984) echoes this notion by pointing out that detailing works that were once the centre of architectural expression, details are now being resolved through drafting technologies. "Real details" were replaced with "virtual" procedures, where details are viewed as production drawings and not as joints or parts of a building. Frampton (1990) summarizes this argument as the reduction of architecture to scenography, or 'value-free aestheticism' (Grassi, 1980).

In addition, the abandonment of a tectonic expression is due in part to an emphasis on space in the perception of architecture (Weiner, 1996). Frampton (1995) observed that man's understanding of space is an integral part of experiencing architecture, and that architecture is always perceived with spatial-temporal emphasis. This observation found roots in the theories of Bruno Zevi (1957) who opined that other arts that contribute to the making of architecture, though important, are always subordinate to the spatial idea. Although space itself does not define architecture, Zevi claims that it embodies the character quality and the essential attributes enclosed by an architectural container. As such, an atectonic approach that underlined artistic and plastic thinking overtook the emphasis of construction in architecture (Garland, 2003). Vitorrio Gregotti (1996) cited the rise of this decorative end as the basis for lack of tectonic expression in architecture. Contemporary architecture employs pastiche as a way to replace the meaning in authentic ornamentation. Construction is viewed as merely a means to achieve spatial effect as an end (Srirangam & Ng, 2012).

In response to the demise of a tectonic expression, Frampton propagated the return to tectonics in his seminal essay Rappel a l'Ordre: The Case for the Tectonic (1990). This reassertion of a tectonic expression was derived from Giogio Grassi's (1980) polemic on the realization of ineffectuality of architecture that trailed in the wake of figurative arts such as Neo-plasticism and Cubism as architecture is a systematic craft in itself, involving established knowledge of construction developed through cultural application and refinement. Frampton defines tectonics, or the 'poetics of construction' (1995), as the art of putting together details through an amplification of its construction within the assembly or architecture of which it is a part (1990). Thus, Frampton suggests the employment of tectonics in the joining of elements as a prospect against the degeneration of cultural consciousness in construction.

Marco Frascari (1984) also emphasized this idea of joining of elements in tectonics. Tectonics is the realization of construction knowledge in the form of 'joints' and contains intrinsic properties that, with the application of detailing in joining parts together, have the potential to construe and give rise to meaningful architectural articulations. He regards details as significant 'joints' in the role they play in constructing and construing meaning as each detail forms a part of a whole, while being wholes in themselves. This paper will further explore this notion of tectonics as 'jointing' from Frascari's view point (1984) by expounding on the definition of 'joints' in relation to tectonics. It also highlights a cultural consciousness in the usage of 'joints' in the Asian and Malaysian context and draws towards the conclusion that 'jointing' can provide a resolution to the production of "meaningful ornamentation wrongly sought after in pastiche", as the amplification of construction can be a source of construal of meaning as a return to the origins of architecture.

Asia has also seen a surfacing of the global phenomenon of homogeneity in mass production and tendencies toward spatial expression. However, a recent awareness of the demise of a tectonic expression in Asia has caused architects in Asia to take redemptive stances in the execution of architecture that is true to cultural consciousness. For example, the works of Malaysian architect, Wooi Lok Kuang. An overview of the works of Wooi Lok Kuang of Wooi Architect reveals a consciousness in the application of construction detailing and materials that stems from a thorough understanding of the climate and the cultural context in which his designs are situated. This paper further explores the design philosophy of Wooi, as well as his works, specifically the Wooi Residence (2004), Ting Residence (2010) and House 9 (2015). These built examples of tectonic expression allude to the entelechy of poetic manifestation of structure within the Malaysian context, displaying material consciousness of an ingrained local building culture that eases the tension between a design intent and it's mode of communication. It analyses these works through the framework of Frascari's notion of tectonics as 'joints', highlighting the constructing and construing of cultural meaning reflected in the presentation and documentation of Wooi's design intent: How can the construction of 'joints' bring about the construing of meaning through tectonics in the works of Wooi Lok Kuang?

2. Materials and Methods

This paper employs a mixed-methodology in designing the research. Literature review, case studies and interview will be used as a method of research and inquiry. It is 'an empirical inquiry that investigates a phenomenon or setting' (Groat & Wang, 2013). The case study will focus on multiple works of Wooi Lok Kuang in order to identify patterns within the design process of Wooi in using tectonic expression to craft meaningful architectural spaces. An analysis of the case study will employ a theoretical framework developed from the review of literature and theoretical writings of various sources. The aim of the case study is to highlight the design thinking behind meaningful constructing and construing of architecture. In addition, Wooi Lok Kuang himself will be interviewed to develop an understanding of his background and his belief system that leads to specific design decisions made.

A review of literature, such as writings, drawings and photographs, will first be carried out to provide a background study on related references pertaining to the notion of tectonics. Citing works from key thinkers such as Kenneth Frampton (1990, 1995), Marco Frascari (1984), Vitorrio Gregotti (1996), and Gottfriend Semper (1989, 2004) provides valid references that can be grouped into common themes, allowing for the research to develop a theoretical framework of the notion of tectonics as constructing and construing.

Included within this research is an interview of Ar. Wooi Lok Kuang. The purpose of the interview is to establish an understanding of the background of the architect, including his upbringing in Malaysia and his education in Australia as well as the formation of his unique design ideologies. The interview with Wooi reveals a correlation between his background, design ideologies and the design decisions undertaken, particularly in 3 of his selected works, the Wooi Residence (2004), Ting Residence (2010) and House 9 (2015).

These selected works will also be analysed through the theoretical framework of 'joints' as the site for constructing and construing of meaning, underlining the design philosophy and methods employed by Wooi in the construal of meaning of each 'joint' or part in relation to its whole. The case studies will attempt to reveal the design thinking and methods adopted by Wooi, whether the design intentions are achieved through conscious or unconscious means. This study attempts to position cultural and contextual consciousness in tectonics as a resolution to homogeneity in the built environment as a result of mass production and spatial emphasis.

3. Findings and Discussion

In Wooi Residence, Ting Residence and House 9, Wooi's careful assembly of 'joints' is highly informed by a pre-determined design concept. Each 'joint' is then situated to articulate a desired spatial experience or quality that reflects the design concept. There is a refinement in the way details are assembled, reflecting an evolution of Wooi's design process with every new project.

An analysis of Wooi's selected works reveals a consistent development in the application of formal and material 'joints'. The three characteristics that drive his tectonic expression, namely, a) Malay cultural influence, b) construction innovation and c) spatial experience, is a culmination of his prior experiences, primarily his childhood in a Malay kampong.

Malay cultural influences can be seen most richly explored in his earliest design, Wooi Residence (2004), which then led to a recurring usage of these 'joints' in his later designs. For instance, the idea of a selang in both Wooi Residence and House 9 provides a formal 'joint' between spaces of varying privacy levels. In addition, the anjung is a recurring material 'joint' seen in both Wooi Residence and House 9. In both instances, although the material 'joint' does not portray an exact replica of detailing, the idea behind the 'joint' is inspired by its original usage in the traditional Malay house. There is also one-off application of 'joints', the tiang seri in Wooi Residence being one such instance.

In most cases, Wooi's construction innovations arise as a response to site forces, or topos. In the case of Wooi and Ting Residences, the form of the building is derived from the existing contours on site. Hence, 'joints' serve to amplify the curved nature of the site, construing a formal link between internal spaces and the external site. Wooi's innovations also stem from a design intent such as Wooi's emphasis on openness in spatial planning to maximize cross ventilation. Material 'joints' employed work towards reducing columns and beams within spaces such as the flat slab system in House 9 that is paired with the use of steel structures.

IOP Conf. Series: Materials Science and Engineering **401** (2018) 012001 doi:10.1088/1757-899X/401/1/012001

Wooi's construal of spatial experiences develops throughout the course of his practice. In his earliest work, Wooi Residence, there exists an experimentation of formal and material 'joints' to create spatial experiences such as the vaulted, leaf-shaped ceiling of the master bedroom or the timber strip walls of the master bath. However, in his later projects, there exists a refinement in Wooi's usage of 'joints' in narrating a journey through the site from the entrance of the house to the sanctuary within. The spiral staircase of Ting Residence formally links the entrance level and the dwelling spaces above with the interplay of materials. In House 9, a similar narrative of a journey from the entrance into the forecourt and later into the internal street shows a clear sequence of formal 'joints' to mediate privacy levels. Wooi's refinement in a systematic methodology of designing from 'joints' has resulted in a unified narrative through the site.

4. Conclusion

The origins of tectonics expounded tectonics as the *poiesis*, or crafting of architecture using constructional knowledge and skills. In Wooi's assembly of formal and material 'joints', he echoes Frampton's and Frascari's school of thought whereby tectonic expression involves technology, or the knowledge of making. For each of his construction innovations, Wooi emphasizes the role of engineers and builders along with their established structural and construction knowledge in the crafting of 'joints'. In the crafting of the main curved roof of Ting Residence, for instance, this formal 'joint' consists of a hierarchy of materials and material 'joints' that is assembled to create a curve that reflects the surrounding *topos*. The result is a tectonic expression that is made possible by a combination of architectural, structural and constructional knowledge and skill in the use of materials and structural systems, a marriage between the technical and the articulatory as opined by Schumacher (2014).

The analysis of Wooi's selected works reveals a consciousness of the culture and context in which his designs are situate. This consciousness is manifested in a consistent construed meaning or intent within his assembly of 'joints'. There exists a tangible and intangible aspect to the meanings being construed in Wooi's usage of 'joints'. Wooi attempts to incorporate reinterpretations of elements of the traditional Malay vernacular within his designs. His use of Malay cultural nuances as formal 'joints' such as the tiang seri as the principal post of Wooi Residence as well as a selang to connect spaces underlines an intentional signification of meaning within his tectonic expression that is enriched by influences found in the Malaysian context.

Wooi's designs begin at 'joints' and junctures, as evidenced by his practice of sketching details and commissioning 1:1 mock-up 'joints' before incorporating them within his designs. Like Frascari opined, Wooi also believes that details or 'joints' are the most elementary units of architecture that is capable of articulating meaning. His constant experimentation with 'joints' stems from cultural and contextual factors. Echoing Frampton's notion of tectonics, Wooi draws inspiration from both the topos and typos to inform the construction of 'joints'.

The origin of Wooi's unique tectonic expression may be traced to the various experiences he underwent. He develops an acute understanding of the Malay culture as well as the Malaysian context. Furthermore, the analysis and synthesis of Wooi's tectonic expression underlines key factors that enriches his design methodology and serves as a remedy to the demise of a tectonic expression in the Malaysian context

Wooi begins each design exploration at the 'joints', respecting the hierarchy of parts to whole. Each formal and material 'joint' draws its inspiration from culture, specifically the Malay culture, as well as context, the topos and typos. Next is the assembly of these 'joints' into a unified tectonic expression that is driven by a design concept. In all three of the cases studied, Wooi emphasizes an openness in spatial planning in order to allow daylighting and cross ventilation to permeate spaces. In all three case studies, Wooi's 'joints' display a systematic assembly of the technical and articulatory that is augmented by drawing from the Malay culture as well as the different characteristics of topos and the various requirements of the client, typos. Wooi utilizes the same design methodology that results in varying manifestations of tectonic expression across each design.

Wooi Residence, Ting Residence and House 9, Wooi's careful assembly of 'joints' is highly informed by a pre-determined design concept. Each 'joint' is then situated to articulate a desired spatial experience or quality that reflects the design concept. There is a refinement in the way details are assembled, reflecting an evolution of Wooi's design process with every new project.

An analysis of Wooi's selected works reveals a consistent development in the application of formal and material 'joints'. The three characteristics that drive his tectonic expression, namely, a) Malay cultural influence, b) construction innovation and c) spatial experience, is a culmination of his prior experiences, primarily his childhood in a Malay kampong.

Malay cultural

Acknowledgements
The authors would like to acknowledge Taylor's University's support of this study in providing funding through the Emerging Research Funding Scheme. The authors would also like to acknowledge the supervisory and research agenda within the Taylor's University Master of Architecture programme.

References

[1] Frampton K 1990 Rappel a l'Ordre: The Case for the Tectonic *Architectural Design* **60** pp 19–25.

[2] Frampton K 1995 *Studies in tectonic culture: the poetics of construction in nineteenth and twentieth century architecture* ed J Cava (Place: Graham Foundation for Advanced Studies & The MIT Press)

[3] Frascari M 1984 The tell-the-tale detail *Theorizing A New Agenda for Architecture: An Anthology of Architectural Theory 1965-1995* ed K Nesbitt (New York: Princeton Architectural Press) pp 500–514

[4] Frascari M 2002 Light, Six-Sided, Paradoxical Fight *Nexus Network Journal* **4(2)** pp 21–37 http://doi.org/10.1007/s00004-002-0031-3

[5] Garland K 2003 Contemporary Architecture and Tectonics *Design and Technology Lectures* ed S. Paris (Barcelona: LISt Lab Laboratorio Internazionale Editoriale) pp 131–134

[6] Grassi G 1980 Avant-Garde and Continuity *OASE Journal for Architecture* **92** pp 17–21

[7] Gregotti V 1996 The Exercise of Detailing *Theorizing A New Agenda for Architecture: An Anthology of Architectural Theory 1965-1995* ed K Nesbitt (New York: Princeton

Architectural Press) pp 496–497)
[8] Groat L & Wang D 2013 Architectural Research Methods. (US: John Wiley & Sons)
[9] Nilsson F 2007 New, Technology, New Tectonics? - On Architectural and Structural Expressions with Digital Tools *Tectonics - Making Meaning* Retrieved from http://publications.lib.chalmers.se/publication/165235%0ANotice:
[10] Schumacher P 2014 Tectonic Articulation – Making Engineering Logics Speak. *Future Details of Architecture.*
[11] Semper G 1989 The Four Elements of Architecture and Other Writings *RES Monographs in Anthropology and Aesthetics* (Cambridge England, New York: Cambridge University Press)
[12] Semper G 2004 *Style in the Technical and Tectonic Arts* ed J. D. Berry & S. Lindberg (Los Angeles: Getty Publications)
[13] Srirangam S & Ng V 2012 Transitions From Then to Now: Tectonic Transformation - WHBC Architects *Architecture Malaysia* pp 72–78
[14] Venturi R, Brown S & Izenour S 1977 *Learning From Las Vegas* (Cambridge, Massachusetts: The MIT Press)
[15] Weiner F H 1996 Towards the Tectonic: A Critique of the Idea of Space in Architectural Education *84th ACSA Annual Meeting* pp 500–504
[16] Zevi B 1957 *Architecture as Space: How to Look at Achitecture* (Horizon Press)

A Study on the Evaluation Tendency of Thermal Sensation in Tropical Region - Targeting short time residents -

K Fukagawa[1], Y Kurazumi[2], A Aruninta[3], J Ishii[4], Y Yamato[5], K Tobita[6], E Kondo[7], A Nyilas[8] and H Oishi[9]

[1,3]Department of Landscape Architecture, Faculty of Architecture, Chulalongkorn University, Bangkok, Thailand, [2,8]Department of Human Environment Design, School of Life Studies Sugiyama Jogakuen University, Nagoya, Japan [4]Department of Architecture, Faculty of Science and Technology, Meijo University, Nagoya, Japan, [5]Department of Architecture and Structural Engineering, Kure National College of Technology, Kure, Japan, [6]College of Sustainable System Sciences Osaka Prefecture University, Sakai, Japan, [7]Department of Creative Engineering, Ariake National College of Technology, Omuta, Japan, [9]JR East Design Corporation, Tokyo, Japan

[1]fukagawak@gmail.com

Abstract. Climate change is getting more and more serious issue in the world. This is known as caused by the increase of carbon dioxide. To stop the increase, many countries agreed to the Paris Agreement in 2015. When to discuss the climate change, temperature drop and increase are focused more often. The big change of the temperature, especially the rise of the temperature, causes the more usage of air conditioning system, and it also causes more energy consumption. In addition, the usage of the air conditioning system is affected by both indoor and outdoor thermal environment. The temperature setting of air conditioning system tends to be lower in the countries situated in tropical region than the countries in the other region. This happens because the thermal sensation of human is easily affected by the outdoor environment. For that reason, to grasp the tendency of thermal sensation in outdoor space is important. In addition, the tendency of short time residence is necessary to wipe out the influence of the thermal environment adjustment by long time living. For that reason, this research targeted the thermal evaluation tendency of short time residents in Bangkok and carried out subject experiments in outdoor spaces to develop landscape design method. As the result of analysis, a different tendency was grasped at one of the measuring points with open sky.

1. Introduction

According to the global climate change, many countries are facing serious problems, especially in the countries suffering the rise of the temperature such as Japan. The rise of temperature causes more deaths by heat stroke in summer in both indoor and outdoor. Furthermore, it also causes more energy consumption and causes the exhaustion of more carbon dioxide.

The death by the heat stroke is caused by losing the balance of body temperature. In addition, the balance of body temperature of human is mainly affected by the heat transfer between human body and the air. However, the thermal sensation of human is also an important factor.

For instance, when the thermal sensation reaches to hot, the human body starts to sweat to control the body temperature, then the body temperature is to be adjusted. For that reason to estimate the thermal sensation is important. However, the thermal sensation is affected by both physical and psychological factors.

There have been many researches targeted physical thermal factors. On the contrary, there are not many research targeting the psychological factors.

Content from this work may be used under the terms of the Creative Commons Attribution 3.0 licence. Any further distribution of this work must maintain attribution to the author(s) and the title of the work, journal citation and DOI.
Published under licence by IOP Publishing Ltd

Matsubara et al. [1] focused on sight sense as psychological effect and had experiment on hue-heat and clarified that the effect of visual stimulus causes 0.3 degree centigrade difference.
Kurazumi et al. [2] studied the effect of artificial and non artificial visual stimulus on human thermal sensation, and mentioned the effect on human's neutral temperature.
Ohno et al. [3] had research on the composite environment of thermal and hue-heat impression, and grasped the interaction on the subject.
Ahmed [4] studied the thermal comfort for the tropical urban environments and showed the importance of development of outdoor comfort model.
According to those researches, it is important to clarify the evaluation tendency on outdoor thermal environment from the psychological aspect, human thermal sensation.
For that reason, the purpose of this research is to clarify the psychological effect on human thermal sensation in outdoor space in tropical region by targeting short time residence to wipe out the influence of environment history.

2. Experiment overview

The research was carried out in Bangkok in rain season. The measurement points were selected in the university premises. 4 points were chosen by considering surrounding environment such as surface ground condition and sky factors. The details of measuring points are shown in Figure 1, 2 and Table 1.

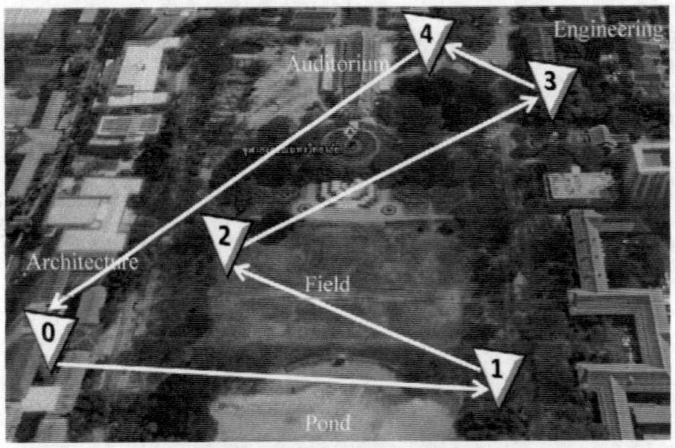

Figure 1. Measuring points

Table 1. Ratio of sky factors

Sky Factor / Measuring Point	Sky	Non Sky	Green
⓪Building court	0.14	0.52	0.34
①Pond side	0.25	0.09	0.66
②North playfield	0.88	0.03	0.09
③Engineering bldg side	0.16	0.28	0.56
④Auditorium plaza	0.42	0.22	0.36

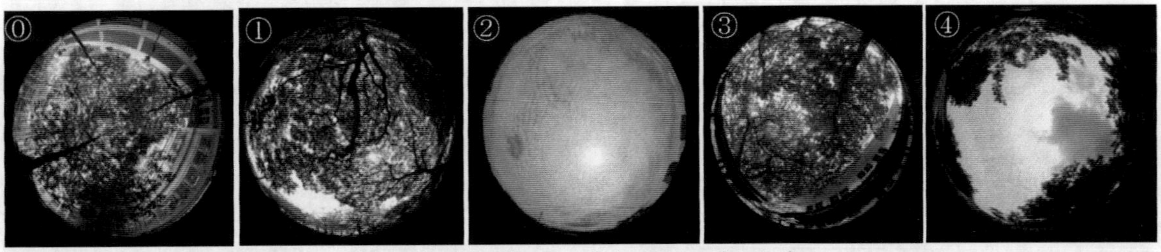

Figure 2. Sky photos of the measuring points

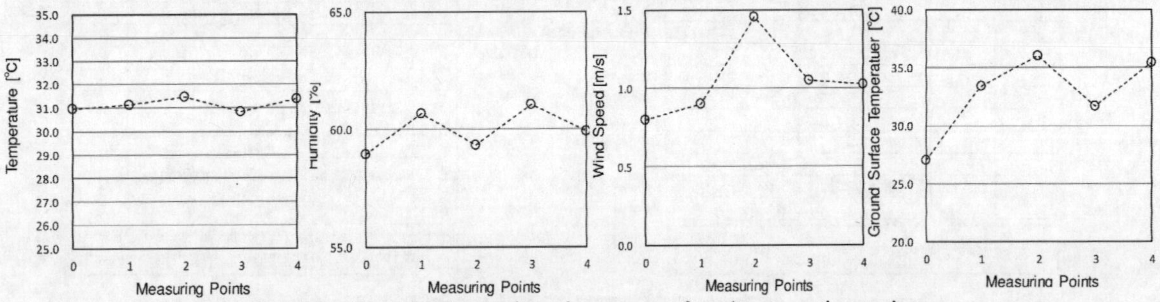

Figure 3. Average thermal environment of each measuring point

The experiment was carried out 4 days with the total subjects of 64 by using declaration vote.
To grasp the thermal conditions of each measuring point, consecutively measured thermal data of 2014 in the same season was used.
The averaged value of temperature, humidity, wind speed, and surface ground temperature is shown in Figure 3. According to the figure, it is clear that there is not big difference of the thermal condition among the measuring points.

3. Details of the declaration vote

The declaration vote was taken for this experiment. The vote is composed of image evaluation part using SD method and thermal sensation part. SD method part contains 30 adjective pairs with 7 level rating scales, by giving the score of +3 to -3. The thermal sensation part is by line rating scale.

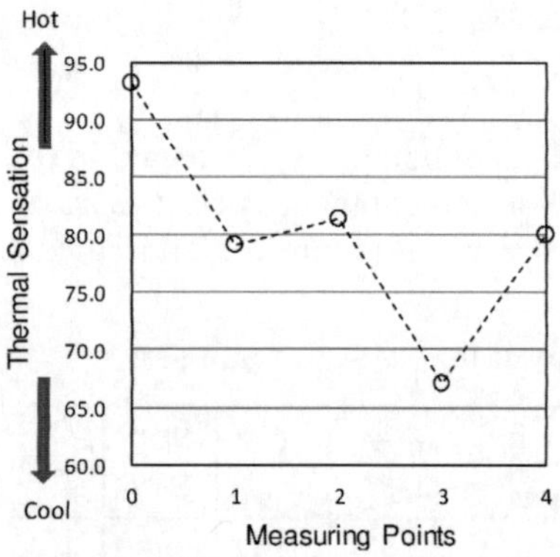

Figure 4. Thermal sensation vote

4. Experiment results

The result of the experiment is shown in Figure 4 and 5. Figure 4 shows the averaged value of thermal sensation in each measuring point, and Figure 5 shows the averaged value of image evaluation result in each measuring point.
As can be seen from Figure 4, the measuring point "0" showed the highest value, then "2". The point "3" showed the lowest value. The measuring point 0 which showed the smallest ratio of sky factor, and the point 3 showed the largest value of green factor as shown in Table 1.
By looking at Figure 5, it can be seen that measuring points "2" and "3" shows different tendency compared to the other points

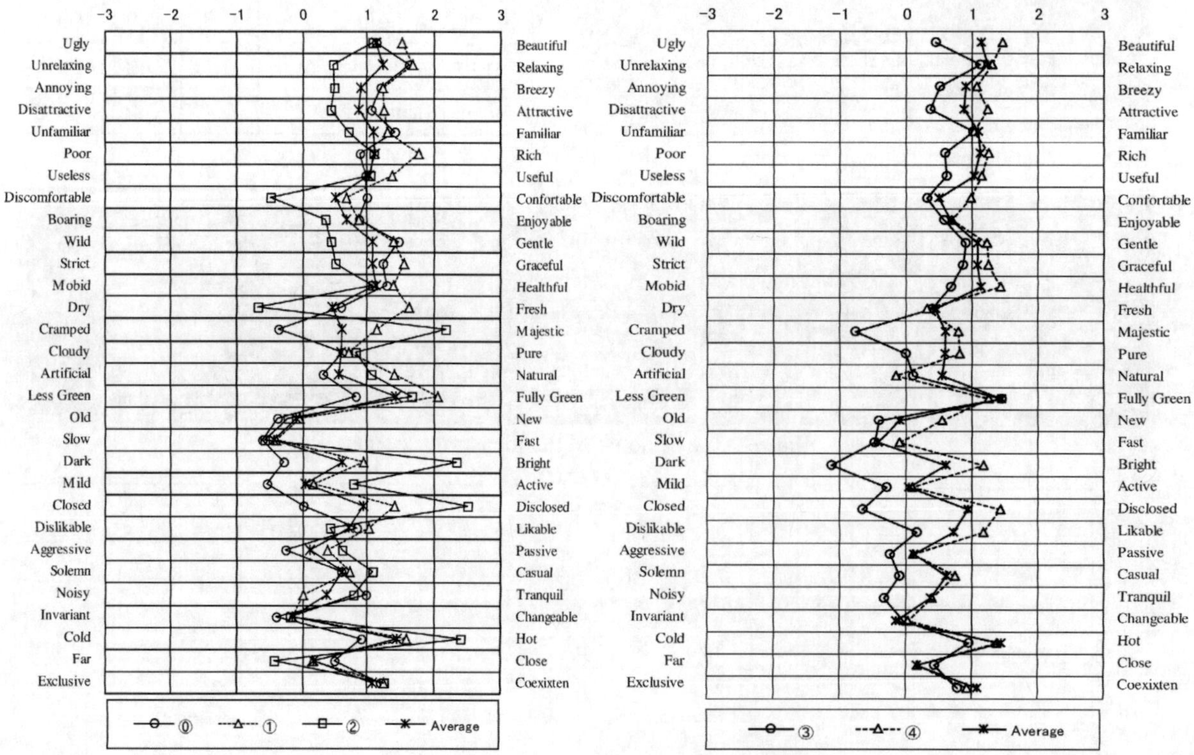

Figure 5. Results of image evaluation

Table 2. Factor loading matrix

Adjective Pairs \ Factor	1st Comfort	2nd Sensation	3rd Impression
Unattractive - Attractive	0.83	0.18	0.23
Strict - Graceful	0.76	0.10	-0.02
Unrelaxing-Relaxing	0.72	-0.12	0.19
Annoying - Breezy	0.72	0.06	0.24
Unfamiliar - Familiar	0.72	0.12	0.19
Ugly - Beautiful	0.71	0.30	0.07
Poor - Rich	0.71	0.31	0.00
Wild - Gentle	0.70	-0.09	-0.17
Dislikable - Likable	0.68	0.25	0.20
Mobid - Healthful	0.67	0.30	-0.06
Discomfortable - Confortable	0.67	-0.10	0.29
Boaring - Enjoyable	0.63	0.19	0.29
Useless - Useful	0.55	0.39	0.19
Cloudy - Pure	0.51	0.40	0.07
Dry - Fresh	0.50	-0.03	0.08
Exclusive - Coexistent	0.42	0.36	0.19
Old - New	0.36	0.20	0.29
Noisy - Tranquil	0.35	0.07	-0.19
Dark - Bright	0.06	0.85	0.10
Closed - Disclosed	0.09	0.84	0.00
Cramped - Majestic	0.17	0.82	-0.12
Mild - Active	-0.05	0.54	0.41
Aggressive - Passive	0.25	0.51	0.36
Cold - Hot	-0.03	0.51	0.00
Solemn - Casual	0.40	0.49	0.15
Less Green - Fully Green	0.37	0.39	-0.24
Slow - Fast	0.16	0.17	0.53
Invariant - Changeable	0.14	0.11	0.46
Far - Close	0.23	-0.23	0.38
Artificial - Natural	0.22	0.30	-0.37

* Factor scores of higher than 0.35 are highlighted.

5. Analysis of the image evaluation structure

To analyse the tendency of the image evaluation results by the subjects, factor analysis was taken. For the factor analysis, varimax rotation method was taken. The most interpret result was obtained when the factor number is set as 3. The result of setting the factor number as 3 is shown in Table 3. The factor score of more than 0.35 are highlighted.

Focusing on the items marked high factor, the first factor is named as "Familiarity" because the items which are relating to comfort, such as "Unattractive - Attractive", "Strict - Graceful" and "Unrelaxing - Relaxing", are categorized.

The second factor is named as "Sensation" because the selected it ems, such as "Dark - Bright", "Closed – Disclosed" and "Cramped -Majestic" which emphasize human sensation are relating.

The third factor is named as "Impression" because the items such as "Aggressive - Passive" "Slow – Fast", and "Invariant – Changeable" are related.

To analyse the tendency of given factor score in each measuring points, the averaged 3 factor scores are plotted in Figure 6.
Looking at the 1st factor, it can be seen that the measuring point "0" and "1" marked high score. In the score of the 2nd factor, measuring point "2" marked the highest, and "3" marked the lowest. In the score of 3rd factor, there is not a big difference among all the measuring points.

6. ANALYSIS ON THE CORRELATION OF THE IMAGE EVALUATION FACOTRS AND THE SKY FACTORS

The result of Pearson's correlation analysis between the image evaluation factors and the sky factors is shown in Table 3.
Only the 1st and the 2nd factor showed the advantage of either 1% or 5% level. The correlation coefficient of the 3rd factor is low comparing with the other factors. In addition, the 2nd factor showed the highest value of correlation coefficient among the 3 factors toward the sky factors.
Focusing on the 1st factor, "Non sky" and "Green" showed the positive value. The 2nd factor, on the other hand, only "Sky" showed the positive value.

7. ANALYSIS ON DEFINITE FACTOR BY MULTIPLE LINEAR REGRESSION

To analyse the definite factor of outdoor thermal sensation, multiple linear regression analysis was taken. For the analysis, factor score was separated to each measuring point and analysed. The result is shown in Figure 7.
Focusing on the multiple correlation coefficient, measuring point 3 showed the highest value and then measuring point "2". The measuring point "4" showed the lowest value.
Focusing on the tendency of standard regression coefficient, only measuring point 2 showed the obviously different tendency. The standard regression coefficient of 2nd factor "Sensation" is negative

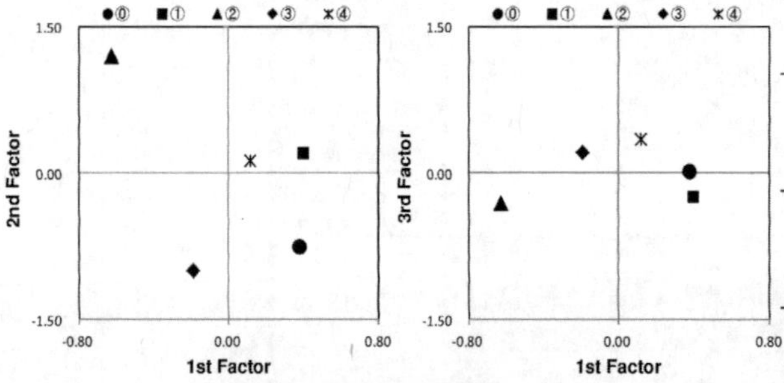

Figure 6. Averaged factor score

Table 3. Result of correlation analysis

Obtained Factors \ Sky Factor		Sky	Non Sky	Green
1st Factor Comfort	Correlation Coefficient	-0.29 **	0.18 **	0.25 **
	Sample Number	296	296	296
2nd Factor Sensation	Correlation Coefficient	0.68 **	-0.60 **	-0.44 **
	Sample Number	296	296	296
3rd Factor Impression	Correlation Coefficient	-0.11	0.12 *	0.04
	Sample Number	296	296	296

**shows advantage at 1% level ** shows Advantage at 5% level

	Measuring Points ⓪	①	②	③	④
Multiple Correlation Coefficient	0.45	0.41	0.49	***0.55***	0.29
Comfort	0.30	0.33	0.29	***0.41***	0.16
Sensation	0.29	0.06	***-0.43***	0.18	0.17
Impression	0.13	-0.17	0.10	***0.19***	0.06

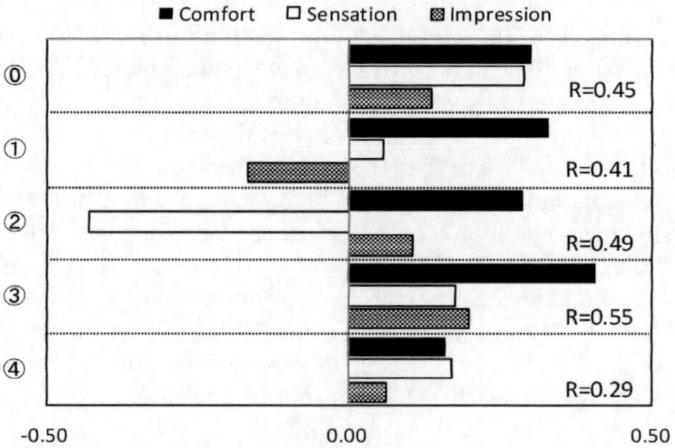

Figure 7. Result of multiple linier regression analysis

value to the thermal sensation. The factor "Sensation" is composed of the items including "Dark - Bright", "Cold - Hot". For that reason, this factor is can be taken as most likely affecting the thermal sensation. In addition, measuring point "2" is with the highest percentage of sky factor which means with more sunlight than the others.
By the deeper analysis of this result, it can be said that the subjects declare the thermal sensation lower than the actual condition of the thermal environment at measuring point "2" where is with larger sky factor and more sunlight.

8. CONCLUSION

In this study, short time resident was targeted to clarify the evaluation tendency of thermal environment. The obtained results are as below.

1) Comparing the results of image evaluation of each measuring point, point "2" showed obviously different tendency.

2) By the factor analysis, 3 factors were obtained. In addition, the averaged factor scores of each measuring point, the measuring point "2" showed the highest value in the 2nd factor. Furthermore, the measuring point showed the lowest value in the 2nd factor.

3) By the correlation analysis between the obtained 3 factors and the sky factors, the 2nd factor showed the highest value of correlation coefficients on all the sky factors.

4) To analyse the definite factor, multiple regression analysis was taken. As the result, the 2nd factor showed the possibility of lower the thermal sensation at the measuring point "2".

In summary of the above results, it is confirmed that by raising the 2nd factor has the effect of alleviating thermal sensation of the short time residents in a space with a high sky factor.

This result can be applied for designing outdoor space targeting short time resident.

ACKNOWLEDGEMENTS

I would like to show my gratitude to whom joined the experiment as the subjects.

REFERENCES

[1] N.Matsubara et al., Kyoto Prefecture University, *The Japanese Psychological Association Conference* pp.570-571, 2004,

[2] Y.Kurazumi et al., Sugiyama Jogakuen University, *Japanese society of Biometeorology Refereed Journal* Vol.48-4 pp.129-144, Psychological effects of the environmental stimuli on thermal sense in outdoor spaces, 2011

[3] H.Ohno H et al., Nagoyaeiyou college, *Architectural Institute of Japan Refereed Journal* No.374 pp. 8-18, The Interactions between Thermal Sensation and Hue-Heat Impression Under Various Thermal Conditions, 1987

[4] K.Ahmed, University of Engineering and Technology of Dhaka, *Energy and Buildings* Vol 35, Issue 1, pp.103-110, Comfort in urban spaces: defining the boundaries of outdoor thermal comfort for the tropical urban environments, 2003

[5] K.Fukagawa et al. Hiroshima International University, *Architectural Institute of Japan Refereed Journal* Vol. 605 pp.91-98, A study on the differences on the formation of the temperature among urban area, rural area and around irrigation ponds in a developing city, 2005

[6] K.Fukagawa et al. Kyushu Sangyo University, Ecology Environment & Conservation International Journal Vol 23, Suppl. Issue pp. S31-S36, A study on the effect of visual stimuli on human thermal sensation – targeting water landscape, 2017

A Study on the Effect of Environment Sound on Human Thermal Sensation

K Fukagawa[1], Y Kurazumi[2], A Aruninta[3], J Ishii[4], Y Yamato[5], K Tobita[6], E Kondo[7], A Nyilas[8] and H Oishi[9]

[1]Department of Architecture, Faculty of Architecture and Civil engineering, Kyushu Sangyo University, Fukuoka, Japan, [2,8]Department of Human Environment Design, School of Life Studies Sugiyama Jogakuen University, Nagoya, Japan, [3]Department of Landscape Architecture, Faculty of Architecture, Chulalongkorn University, Bangkok, Thailand, [4]Department of Architecture, Faculty of Science and Technology, Meijo University, Nagoya, Japan, [5]Department of Architecture and Structural Engineering, Kure National College of Technology, Kure, Japan, [6]College of Sustainable System Sciences Osaka Prefecture University, Sakai, Japan, [7]Department of Creative Engineering, Ariake National College of Technology, Omuta, Japan, [9]JR East Design Corporation, Tokyo, Japan

[1]fukagawak@gmail.com

Abstract. In recent years, there have been several devastations caused by the global climate change. Global climate change is considered as relating to the global warming and is becoming a serious problem. Global warming is related to the increase of carbon dioxide. The increase of the carbon dioxide is caused by the energy consumption increase. In addition, air conditioning dominates the large ratio of the energy consumption. Furthermore, according to the huge earthquake happened on March 11, 2011 in Japan, the country's energy policy is facing a turning point, which is to establish less harmful and less energy based society than ever. For that reason, to decrease the energy consumption, how to reduce the energy consumption by the air-conditioning is an important issue. There have been several research relating to the topic. However, those researches did not study on the effect of the environment sound on the human thermal sensation. For that reason, the purpose of this research is to clarify the effect of environment sound on human thermal sensation. As the result of this study, the environment sounds, such as "Rain" "Stream" "Water Bamboo" showed the effect to lower the human thermal sensation.

1. Introduction

Climate change is getting more and more serious issue in the world. The change is known as caused by increase of GW gasses, such as carbon dioxide which is the largest exhaust by human activity relating to the energy consumption in the world. In addition, the energy usage of buildings shows the largest ratio in the consumption.

According to the fact that most of energy consumption is caused by the buildings equipment, it is important to reconsider the equipment.

Focusing on the building equipment, air conditioning shows the large ratio. For that reason, it is important to reduce the energy consumption by air conditioning system.

To reduce the energy consumption, there are many research targeted the effect of water surface, hue-heat, and floor cooling system as followings.

Fukagawa et al. [1] carried out a research targeting the effect of irrigation pond on surrounding thermal environment by long term outdoor measurement of thermal environment. This research grasped the effect of irrigation pond to lower the surrounding temperature, about 1.5 degree centigrade as the maximum.

Content from this work may be used under the terms of the Creative Commons Attribution 3.0 licence. Any further distribution of this work must maintain attribution to the author(s) and the title of the work, journal citation and DOI.
Published under licence by IOP Publishing Ltd

In addition, Matsubara et al [2] carried out a research targeting hue-heat by experiments and clarified that the effect of visual stimulus causes 0.3 degree centigrade difference. Furthermore, to clarify the effect of artificial and non-artificial factors as visual stimulus on human thermal sensation, Kurazumi et al [3] had an experiment and grasped the tendency.
There are also the researches on the composite environment of thermal and hue-heat impression in the several room temperatures. Ohno et al [4] tried to grasp the interaction of the subject by experiment.
Fukagawa et al. showed that the effect of water on the temperature is smaller than how the human feels and detects. In addition, the result of Matsubara et al. showed that the human's thermal sensation can be affected by the visual stimulus, such as color. Furthermore, Kurazumi et al. clarified the difference of the effect by artificial and non-artificial visual stimulus on human's neutral temperature.
Hishida et al. and Ohno et al. grasped the interaction of the composite environment. According to the results, there might be some possibilities that the effect of water landscape can be not only as physical effect on surrounding temperature but also as the effect on human's thermal sensation.
For that reason, the purpose of this research is to clarify the effect of sound on human thermal sensation.

2. Experiment overview

This research is studied by a subject experiment. The experiment was carried out in the artificial weather chamber in Kyushu Sangyo University in Japan, in summer in 2014. For the experiment 6 male students were selected as subject. They were age of 20 and 21, and were asked to wear prepared unified clothes, which are short sleeve T-shirt and short pants made of 100% cotton.
During the experiment, subjects were asked to sit on a chair calmly and asked to fill in the declaration vote after environment sound was played for approximately 5 minutes. Then asked to fill in the declaration vote which is composed of SD method part and thermal sensation part.
Totally of 9 different sound conditions, including no environment sound, were set for this experiment. The details of sounds including the sound level is shown in Table 1. The 30 adjective pairs used for SD method is shown in Table 2.
This experiment targeted slightly uncomfortable thermal condition. For that reason, temperatures set for the experiment were 28 and 30 degree centigrade. The moisture was controlled to be between 50 to 60%.
The results of the evaluations are averaged in each temperature setting and the sound type, such as water-related and non- water related sound, are shown in Fig.1 and 2.
By looking at Fig. 1, which is the result of 30 degree centigrade setting, it is obvious that the water-related environment sounds marked higher than the other non-related sounds. In addition, the items which are "Fresh - Dry" and "Cool - Hot" showed big differences.
By looking at Fig.3, which is the result of 28 degree centigrade setting, there is not obvious tendency

Table 1. Utilized Environment Sounds

Type of Environment Sound	Inndication	dB
Background Noise	(BG)	-
Water Filled Bamboo	(WB)	66.0
Stream	(ST)	60.0
Rain	(RA)	59.0
Big Wave	(BW)	68.0
Water Fall	(WF)	60.0
Tree Sound	(TS)	55.0
Tayphoon	(TY)	61.0
Crow Cry	(CC)	58.0

Table 2. Adjective Pairs for SD Method

Enjoyable	-	Boaring	Attractive	-	Dis-attractive
Crisp	-	Gloomy	Fast	-	Slow
Beautiful	-	Ugly	Airy	-	Heavy
Open	-	Closed	Active	-	Inactive
Bright	-	Dark	New	-	Old
Pure	-	Cloudy	Cold	-	Hot
Near	-	Far	Changeable	-	Fixed
Necessary	-	Unnecessary	Natural	-	Artificial
Rich	-	Poor	Healthy	-	Ill
Energetic	-	Docile	Noble	-	Cramped
Gentle	-	Tight	Calm	-	Rough
Comfortable	-	Discomfortable	coexistensive	-	Exclusive
Big	-	Small	Fresh	-	Dry
Familiar	-	Unfamiliar	Quiet	-	Bustling
Likable	-	Dislikable	Masculine	-	Feminine

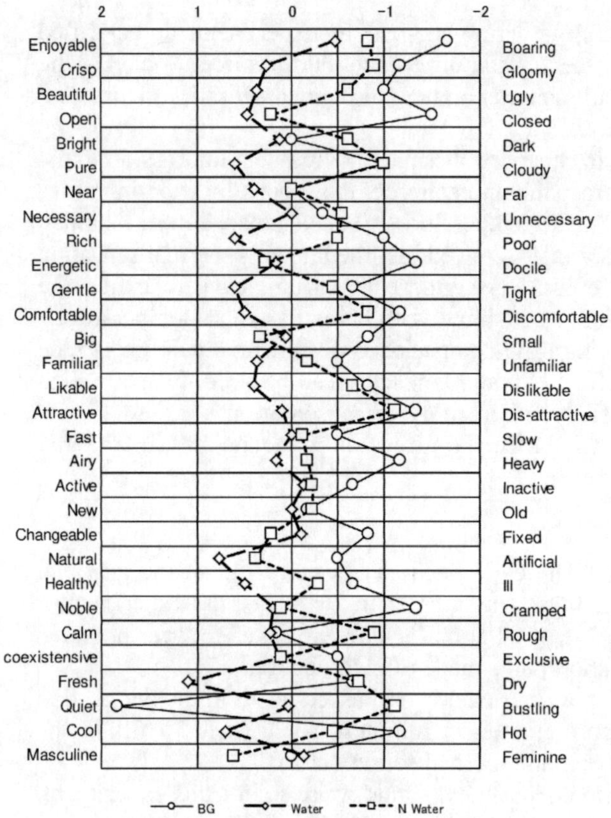

Figure 1. Average Score of Image Evaluation at 30 degree centigrade

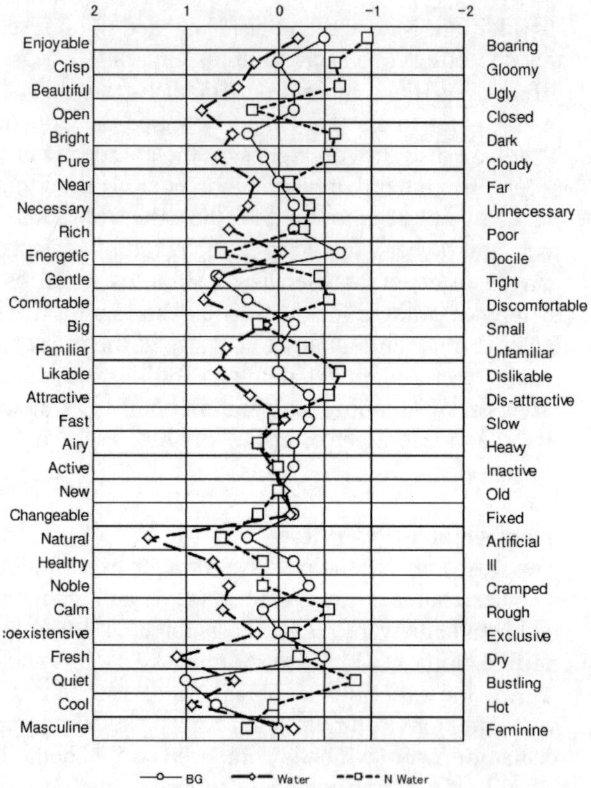

Figure 2. Average Score of Image Evaluation at 28 degree centigrade

as 30 degree setting. However, the items which are "Fresh - Dry" and "Cool - Hot" showed little bigger differences than the other items.

3. Analysis of evaluation structure

To analyze the evaluation structure of the subjects, factor analysis was taken. "Bartlet method" was utilized for the analysis.

For this analysis, the results of all the environment sounds and all the temperature settings were combined and were analyzed. The result of the analysis, the factor loading matrix, is shown in Table 3.

First, the analysis was done by setting the eigenvalue as more than 1.0 and obtained 5 factors which could not clearly be identified. Then changed the setting of factor number down from 5 to 2. When the factor number is set as 3, clearly identified factors were obtained.

The first factor is named as "Emotion" because the items such as "Pure", "Comfortable" and "Crisp" which are relating to human emotion.

The second factor is named as "Movement" because the items, such as "Fast", "Energetic" and "Active" are categorized.

The third factor is named as "Space" because the items such as "Open" "Coexistence", and "Noble" are categorized.

4. Analysis of average factor score

The point diagrams of average factor score in each environment sound differed in each temperature setting are shown in Fig. 3 and 4.

IOP Conf. Series: Materials Science and Engineering **401** (2018) 012003 doi:10.1088/1757-899X/401/1/012003

Table 3. Result of Factor Analysis

Factor / Adjective	1st Factor	2nd Factor	3rd Factor
	Emotion	Movement	Space
Pure	0.881	-0.172	0.214
Comfortable	0.873	-0.136	0.239
Crisp	0.857	-0.110	0.174
Beautiful	0.854	-0.114	0.212
Gentle	0.839	-0.261	0.047
Likable	0.816	-0.071	0.268
Necessary	0.744	0.080	0.141
Attractive	0.742	-0.037	0.299
Cool	0.740	0.165	0.076
Calm	0.728	-0.436	0.174
Bright	0.705	-0.028	0.116
Fresh	0.675	0.014	0.199
Quiet	0.670	-0.569	-0.100
Healthy	0.670	0.086	0.525
Masculine	-0.665	0.130	0.094
Rich	0.661	0.097	0.356
Enjoyable	0.588	0.265	0.169
Airy	0.563	0.236	0.059
Familiar	0.533	-0.090	0.457
Natural	0.524	0.281	0.374
Near	0.155	0.078	0.018
Fast	0.017	0.684	-0.357
Energetic	-0.353	0.671	0.171
Big	-0.200	0.665	0.064
Active	0.125	0.582	0.134
New	0.184	0.282	0.009
Open	0.460	0.090	0.671
Coexistensive	0.326	-0.175	0.624
Noble	0.392	0.372	0.539
Changeable	-0.108	0.091	0.488

Looking at Fig. 3 which is 30 degree centigrade setting, by focusing on the 1st factor, 4 of the 5 water related sounds marked positive score. Non-water related sounds, on the other hand, marked negative score. Beside the 1st factor, on the 2nd and the 3rd factor, there was no clear tendency.
Looking at Fig. 4 which is 28 degree centigrade setting, by focusing on the 1st, all the water related sounds showed the positive score. However, no sound also showed the positive score, which is close to water fall sound. Furthermore, the 2nd factor and the 3rd factor did not show the obvious tendency as the 1st factor.
By the analysis of the average factor score in each sound, some tendencies are confirmed on the 1st and the 2nd factor.

5. Analysis on the correlation of the factors and thermal sensation vote

IOP Conf. Series: Materials Science and Engineering **401** (2018) 012003 doi:10.1088/1757-899X/401/1/012003

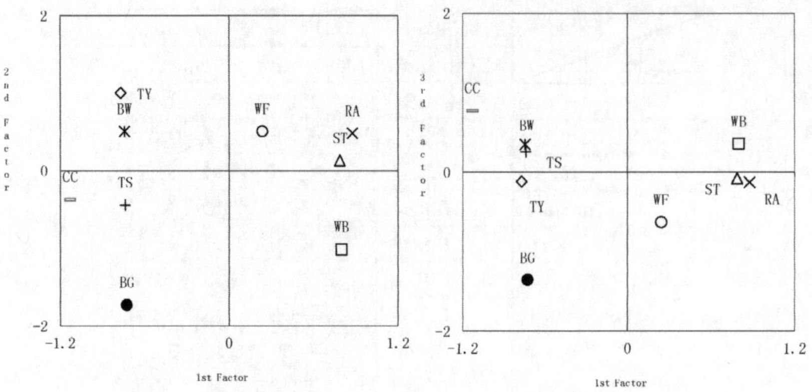

Figure 3. Point Diagrams of Average Factor Score at 30 degree

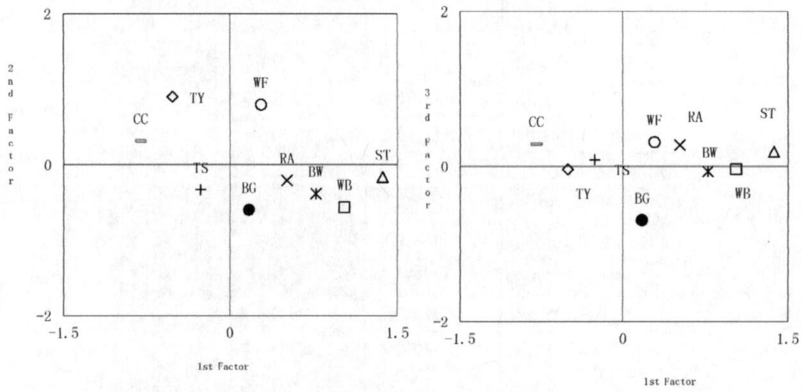

Figure 4. Point Diagrams of Average Factor Score at 28 degree

Table 4. Correlation of the Factors and Thermal Sensation Vote

		1st Factor Emotion	2nd Factor Movement	3rd Factor Space
Thermal Sensation	Correlation Coefficient	-.397**	-.315**	-.048
	Significance Probability	.000	.001	.623
	Number	108	108	108

** shows the 1% significance coefficient of correlation

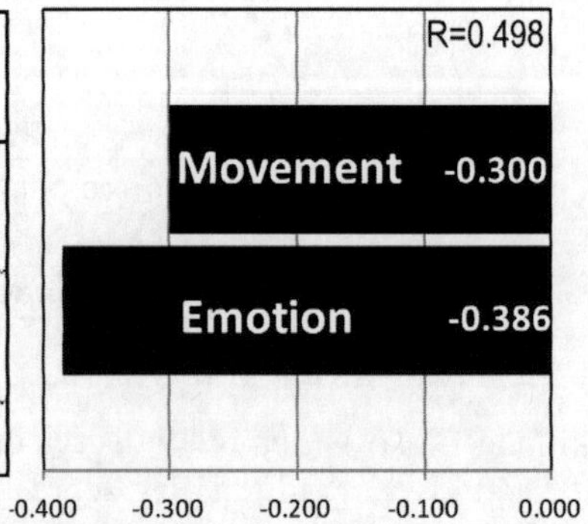

Figure 5. Result of Multiple Linear Regression Analysis

To analyse the correlation among the 3 factors and thermal sensation vote, Pearson's correlation analysis was taken. The result is shown in Table 4.

By focusing on thermal sensation vote, “Emotion” and “Movement” showed significance coefficient of correlation of 1%. In addition, “Emotion” showed the bigger value than “Movement”.

According to the results of the previous chapters, it is possible to say that the essences which are evaluated as “Emotion” by the subjects affect the thermal sensation of the humans. Furthermore, the effect appears both in summer and winter.

6. Definite factor analysis on thermal sensation vote

To analyze the definite factor on thermal sensation vote, multiple linear regression analysis was taken. For this analysis, 2 factors, which showed the 1% significance coefficient of correlation by the Pearson's correlation analysis, were utilized.
The score of the thermal sensation was set as dependent variable. The scores of the 2 factors were set as independent variables. The result of multiple linear regression analysis is shown in Figure 6.
The multiple correlation coefficients by the analysis showed 0.498 which can be taken as high reliability by considering the fact that the thermal sensation is in step scale.
The both factors showed negative correlation to the thermal sensation. This means when the factor score increases, the subjects feel that the thermal condition is cooler.

7. CONCLUSION

By this research, it can be said that the environment sounds which mark high score in the 1st factor affect the thermal sensation more than the other sounds. In addition, the 3rd factor does not clearly affect the thermal sensation.
The obtained 4 results are as follows.

1. By the factor analysis 3 factors were obtained from the environment sounds.
2. 2 of the 3 factors showed negative affect with the significant coefficient correlation on the thermal sensation and the 1st factor showed the largest ratio. This tendency is more obvious under 30 degree centigrade setting than 28.
3. Among all the environment sounds, "Rain" "Stream" "Water Bamboo" marked the large score on the 1st factor
4. The sounds mentioned above have possibility to lower the human thermal sensation and increase the thermal comfort. However, the tendency is more obvious at 30 degree centigrade which means that the sound is more effective on the slightly hot thermal condition.

Acknowledgements

I would like to show my gratitude to whom joined the experiment as the subjects.

References

[1] K.Fukagawa et al. Hiroshima International University, *Architectural Institute of Japan Refereed Journal* Vol. 605 pp.91-98, A study on the differences on the formation of the temperature among urban area, rural area and around irrigation ponds in a developing city, 2005
[2] N.Matsubara et al., Kyoto Prefecture University, *The Japanese Psychological Association Conference* pp.570-571, 2004,
[3] Y.Kurazumi et al., Sugiyama Jogakuen University, *Japanese society of Biometeorology Refereed Journal* Vol.48-4 pp.129-144, Psychological effects of the environmental stimuli on thermal sense in outdoor spaces, 2011
[4] H.Ohno H et al., Nagoyaeiyou college, *Architectural Institute of Japan Refereed Journal* No.374 pp.8-18, The Interactions between Thermal Sensation and Hue-Heat Impression Under Various Thermal Conditions, 1987
[5] K.Ahmed, University of Engineering and Technology of Dhaka, *Energy and Buildings* Vol 35, Issue 1, pp.103-110, Comfort in urban spaces: defining the boundaries of outdoor thermal comfort for the tropical urban environments, 2003
[6] K.Fukagawa et al. Kyushu Sangyo University, Ecology Environment & Conservation International Journal Vol 23, Suppl. Issue pp. S31-S36, A study on the effect of visual stimuli on human thermal sensation – targeting water landscape, 2017

THE APPLICATION OF TRADITIONAL ARCHITECTURE AS PASSIVE DESIGN STRATEGIES FOR MODERN ARCHITECTURE IN HOT DRY CLIMATE

Irma Handayani Lubis[1], Mochamad Donny Koerniawan[1], and Rachmawan Budiarto[2]

[1]Department of Architecture Institut Teknologi Bandung, Bandung, Indonesia
[2]Department of Nuclear and Physics Engineering Universitas Gadjahmada, Yogyakarta, Indonesia

Email: lubis.irma.handayani@gmail.com

Abstract. Traditional Architecture in Cairo is noticeable to give a great respond to climatic conditions in hot dry climates. This paper aims to analyse the application of traditional architecture elements as passive design strategies in modern architecture of Cairo. The methodology of this research is based on a computer simulation. As a case study of one building mass was design and simulated in this study. Building design was based on the application of traditional residential building in Cairo. Then, the performance of building design was simulated in terms of solar simulation, wind direction, and also heating and cooling loads. From the result, the application of traditional element was relatively effective as passive design strategies to reduce solar heat gains and applying natural ventilation because it has less opening in the south, catch the breeze air from south-west and north-east axis, while consumed less heating load. However, the cooling load was relatively the same with the conventional building, this might be lowered by using another passive design strategies. *Keywords-traditional; modern; architecture; cairo; hot dry climate; climatic; design.*

1. Introduction

1.1. Climatic Condition

Egypt is a hot-arid climate country that is located in 30°2′N 31°13′E. The main characteristic of hot-dry climate is the extreme temperature, where during the day the temperature is extremely high while during the night the temperature is relatively low. According to World Meteorological Organization (UN) from 1971 to 2000, the highest temperature was 47.8˚C, and the lowest temperature was 1.2˚C. The number of rainfall per year in Cairo is extremely low. The precipitation reached 0 mm during summer (July, August, Sept) [1]. The average relative humidity is 56%, with minimum 46% and maximum 61% [2]. In terms of amount of sunshine hours, Cairo received an excessive number during summer, with maximum 363 hours in July [1].

Content from this work may be used under the terms of the Creative Commons Attribution 3.0 licence. Any further distribution of this work must maintain attribution to the author(s) and the title of the work, journal citation and DOI.
Published under licence by IOP Publishing Ltd

1.2. Traditional Architecture in Cairo

Traditional architecture in Cairo or identified as Islamic Cairo are basically classified as four different periods, Fatimid (from 969), Ayyubids (1171-1250), Mamluks (1250-1517), and Ottoman period (1517-1805) [3]. Architecture in Fatimid period recognized by the development of Mosque (Mosque of Ibn Tulun, Mosque of El-Hakim, and Mosque of Al-Aqmar) which is characterized by square plan, large inner courtyard in the centre, arcades, fountain, minaret, and also decorative crave pattern. Ayyubids period known as the development of Islamic school (madrasah) building (Al-Salih Nagm Al-Din Ayyub) where the square plan and Minaret was the remaining element in this period. This era was the transition between the late Fatimid and the early Mamluk. Mamluk's period combined mosque and madrasah as one unit block. This period was recognized as the development of giant buildings where mosques become monumental. Building also characterized by geometric decorative pattern.Cairene homes were developed from late Mamluks and throughout Ottoman period. Some element in Cairene homes are a great application of passive design strategies. Sahn (inner courtyard), malkaf (wind catcher), and takhtaboosh (courtyard arcade) utilize as natural ventilation. Mashrabiyya (wooden lattice screens), maq'ad (veranda), shesh (overhang), shuksheika (skylight) are applied as passive solar design. While, nafora (fountain), and salsabil (water cooling plate) are implemented as evaporative cooling strategy.

Hot dry climate is characterized by its excessive sunlight, dry, and low precipitation conditions. The extreme climatic conditions force traditional residential in Cairo to be adaptable to achieve comfort. Passive design strategies were applied as a respond to climatic conditions.

Mohamed (2014) said that traditional houses in Cairo were built close each other to confront climatic conditions [4]. According to Olgyay [10], the arrangement of buildings shelter design in hot-arid climate can be managed by house types, general arrangement, plan, form and volume, orientation, interior, and colour.

Fathy (1986) stated that increasing the height of the building provide shading to indoor house spaces and use as shelter from direct sunglight [8]. Indirect entrance in the Cairene house has two turning, where the first turning protect inside house from wind, dust, noise, and the harshness from outside [7] and the second turning deliver to the inner court.

Inner courtyard in Cairene houses were utilized as ventilation, daylight, and interaction among building masses [5]. Gianni (1988) stated that walls and floors of the courtyard stored solar energy during the day while releasing hot air and storing cold air during the night [9]. Hakim (1988) argued that thick walls and inner courtyards of residential buildings serve as noise barrier and shading element [6].

Fathy (1986) said *qa'a* between two North-South axis *iwans* use as passive solar energy in the summer and winter [8]. *Mashrabiyya* provides lighting and airflow [8] while preventing dazzle [4]. Mohamed (2014) commented that *shokhsheskhas* utilize to enhance daylight and reservoir to release the hot air and replace it with the cool air [4]. He also added that *malqafs* serve as ventilation traps to capture the cool air from outside [4].

This paper aims to analyse the implementation the traditional elements as passive design strategies in modern building in hot dry climates. This study also investigate the performance of the building by simulate the solar radiation, shading pattern, wind direction, ventilation, and also heating and cooling loads.

2. Methodology

Hot and dry climate is an extreme condition where people feel uncomfortable both during day and night. People can be suffered from the hot radiation and sunlight during the day, while endure the cool breeze during the night. Therefore, this type of climate is challenging to be exercised.

Cairo was selected as a case study for this research because it represents the era of traditional architecture in hot dry climate. The building site was selected close to the Nile River to minimize hot and dry while optimizing the wind and moisture for cooling strategy. The building mass and orientation was assumed to be given from the client.

This research used no data collection where the data is purely based on the climatic data (sun path, wind speed, and wind direction) from Revit. The analysis data based on the Revit simulation result.

There were three phases of simulation, solar simulation, wind simulation, and energy performance simulation. Solar simulation was conducted based on summer (20th June) and winter (21th December) time which was simulated every 4 hours, from 06.00 am, 10.00 am, 14.00 am, and 18.00 am. Wind and energy performance simulation was simulated annually.

The analysis started with the solar simulation which illustrated the annual sun path diagram and shading pattern of the building mass, then the wind simulation which produced the annual wind rose and wind flow inside the building, and lastly the energy simulation which described the heating and cooling load of the building.

3. Results and Discussions

3.1. Design Solution

The design concept of the building was creating an adaptable building which has less energy consumption by the application of passive design through traditional elements. This strategy used to confront extreme climate condition, which is hot and arid.

The building was designed with some strategies that can accommodate the occupant comfort. This building utilizes some strategies, such as natural ventilation, passive solar heating, and cooling evaporation which are adaptable whether in maximum or minimum temperature. In the case of facade, the role of large opening windows and sun shading are essential for achieving thermal comfort. Regarding with passive solar heating, walls are thicker in purpose to store the heat and reject sun radiation going through the building. Cooling strategy is applied in form of green roof design. The roof resembles the role of inner courtyard (see Figure 1).

The concept building plan is open space. The absence of room divider contributes to circulate the fresh air in all over indoor area. Each floor is linked by the usage of voids. Voids are located in some spot in second floor. The purpose of this arrangement is to create the high-quality of air flow (see Figure 2 and Figure 3).

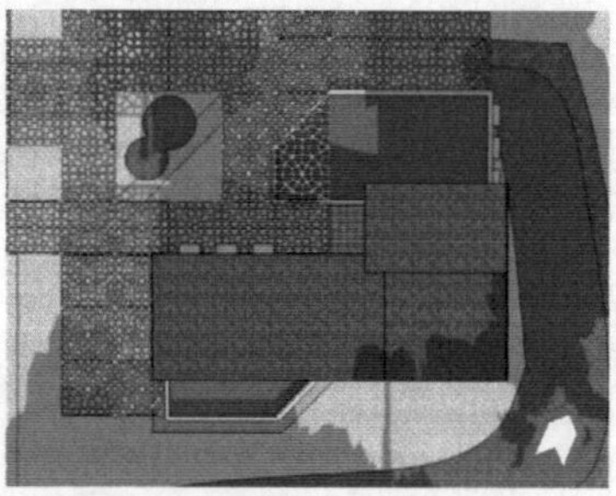

Figure 1. Site plan of one block building design

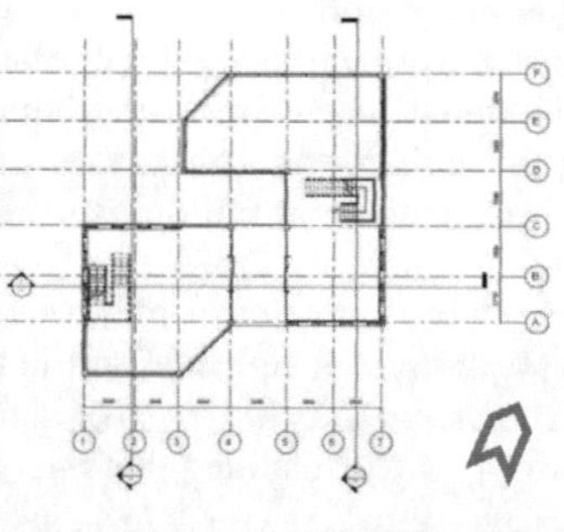

Figure 2. 1st Floor plan of the building

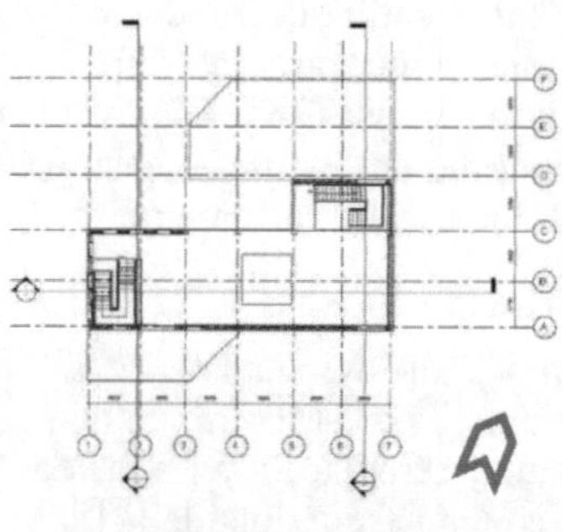

Figure 3. 2nd Floor plan of the building

The arrangement of building facades aims not only as aesthetics but also as tool for passive solar strategy. The thick of the walls is 30cm in purpose to collect the heat energy during winter. Meanwhile during summer, the walls reduce the solar radiation which can raise the air temperature inside the building. Large shading devices (mashrabiyya) are attached in west façades to reduce summer heat and dazzle. This is because the west side is the most sun-exposed part during summer. *Mashrabiyya* is also applied to affirm the characteristic of Cairene house. The Islamic pattern of *mashrabiyya* is applied in some facade elements, such as windows, doors, and railing.

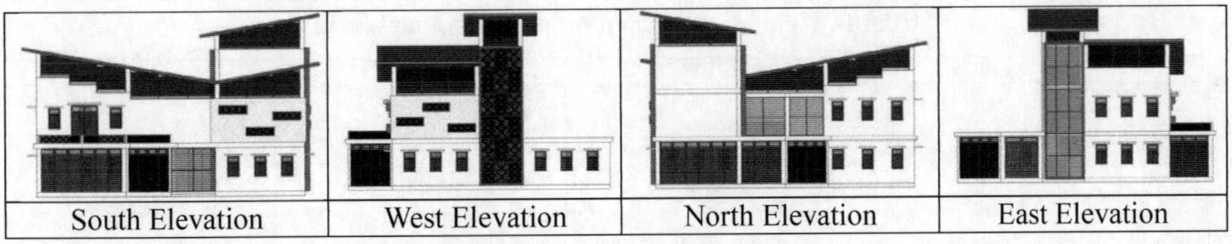

Figure 4. Elevation of the building

As depicted in Figure 5, the position of windows was sink in the building surface, meanwhile mashrabiyya is located far in front of the windows to protect the excessive direct sunlight entering the indoor space. Double frame windows (qamariyyah) are installed in building facades (see Figure 6). Overhang in the upper place serve to shade the windows. The function of the first frame with tight latticework is to reduce direct sunlight during the day, and as an inlet for wind circulation.

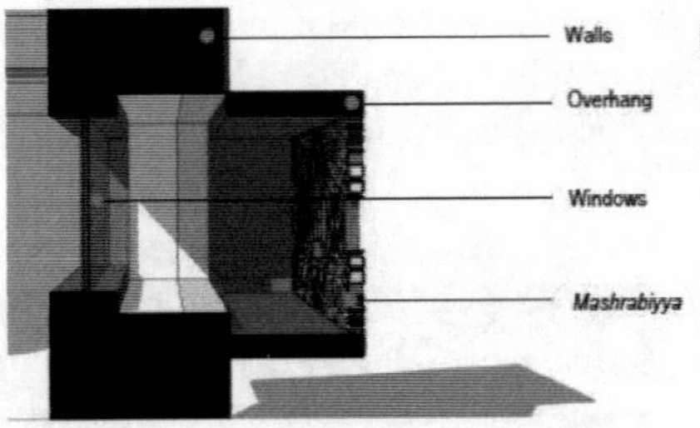

Figure 5. Detail Section of Mashrabiyya

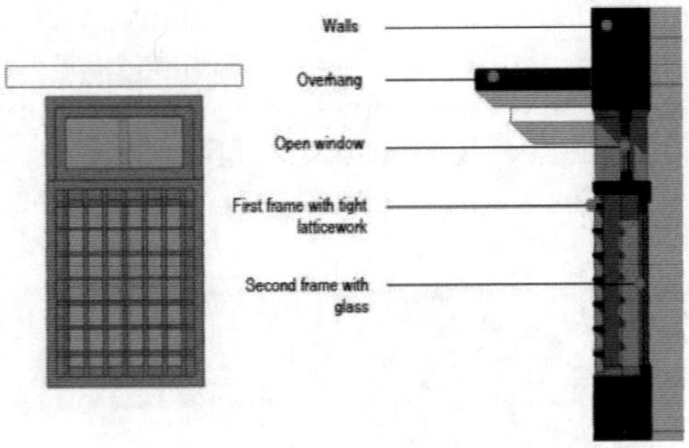

Figure 6. Detail Section of Opening Windows (qamariyyah)

3.2. Solar Simulation (Summer & Winter)

Cairo is located in north hemisphere thus the sun position dominantly comes from the south. This can be used as passive solar design to applied shading device in the south area to reduce solar heat gains during summer and optimize the solar heat gains in the south area during winter.

It can be seen from Table 1 the daylight and shading pattern of the building. Daylight hours during spring and summer time is 13 hours (from 06.00 to 19.00), while during fall and winter is 10 hours (from 07.00 to 17.00). It is also noticeable that the most exposing-sun area is in the north-south axis, while the west-east axis is fully shaded during summer. During winter, south area is unshaded because shading area is only located in west, north, and east area. This strategy can be useful for passive solar strategy during summer and winter.

Table 1 The roof thermal properties of the house

	06.00	10.00	14.00	18.00
SUMMER	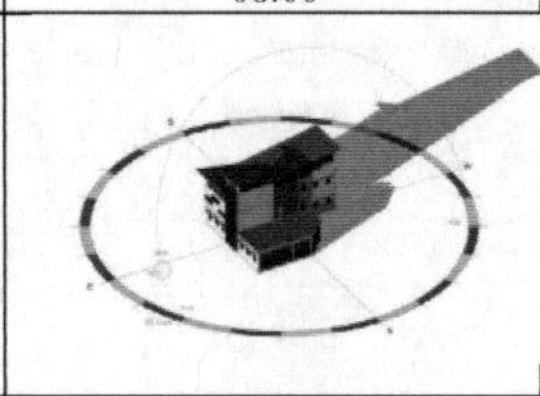	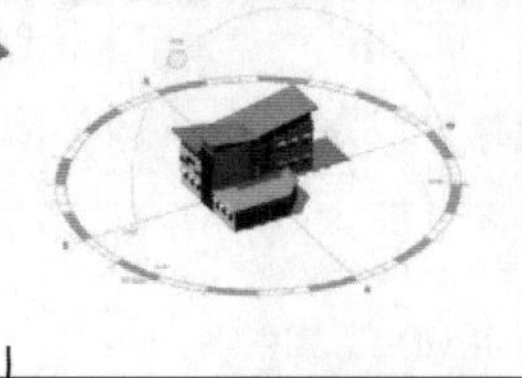	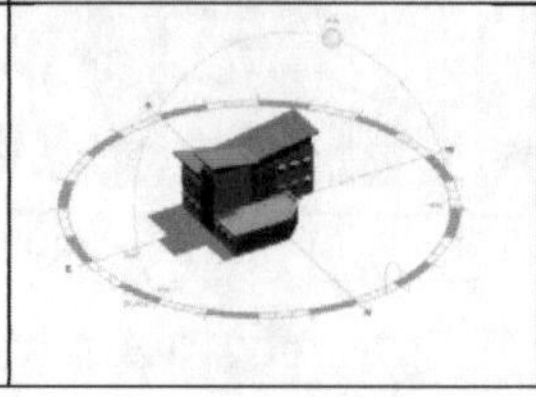	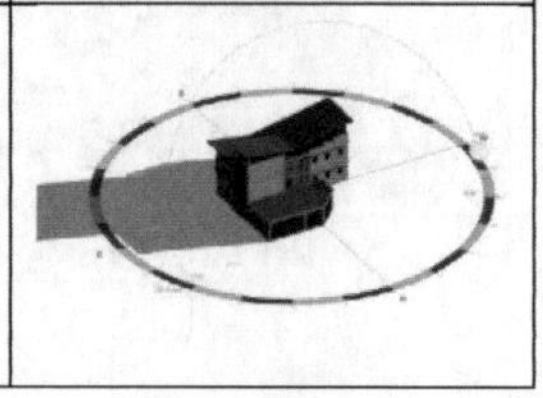
WINTER	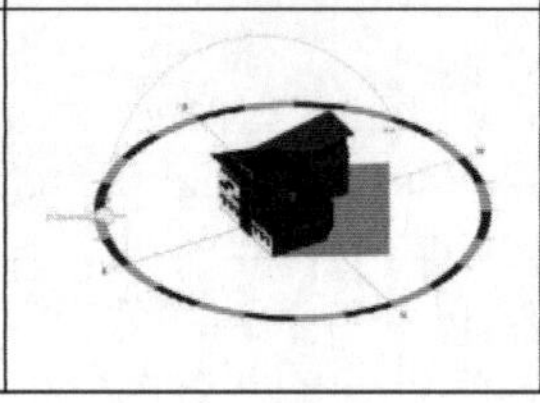	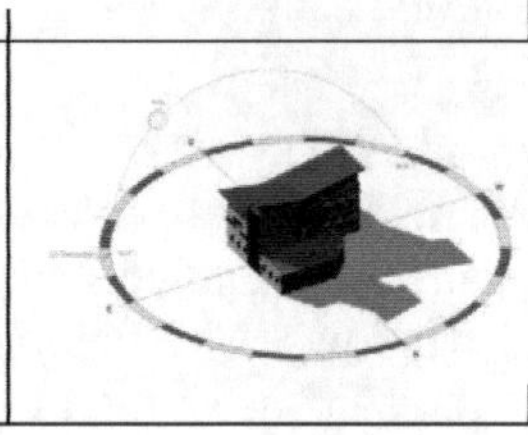	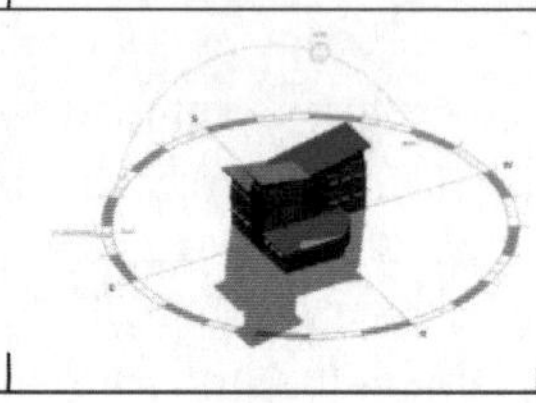	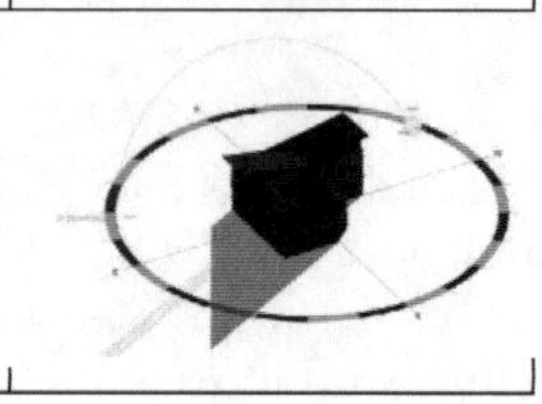

3.3. Wind Direction

From the energy analysis data in Revit simulation (see Figure 7) the prevailing wind comes mostly from the north east part with 35 km/ hours. This is because the site selected is alongside with Nile River in the east (see Figure 8), so it gives impact to the prevailing wind. Located near the Nile River would increase the humidity through evaporation. It also provides the advantage not only offer attractive scenery but also for natural ventilation in the building.

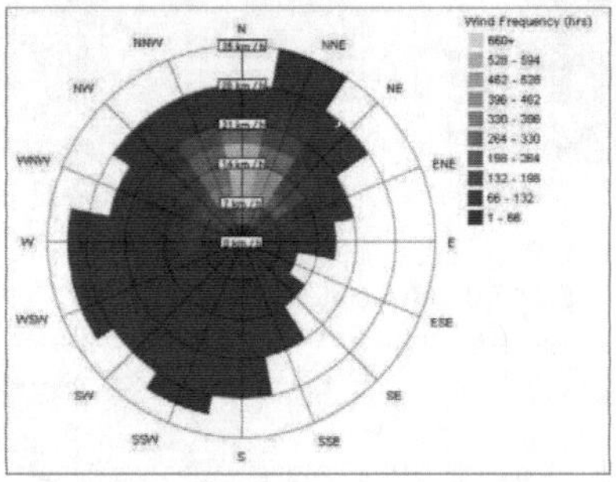

Figure 7. The annual wind rose, frequency direction (Revit simulation)

Figure 8. the site selected is alongside with Nile River

3.4. Natural Ventilation

The natural ventilation in the building was designed by means of stack effect and cross ventilation. Stack effect came from the application solar chimney. This solar chimney serves as reservoir, to release the warm air during the day and replace it with fresh air during the night. Large opening windows the first and second floor provide the cool air and shove the stale air to escape through opening in the tower (see Figure 9). Cross air circulation also occurred inside the building through the addition of void. The function of void in the center allows the air circulation coming from the lower opening to the upper opening (see Figure 10).

Figure 9. Wind analysis for section 1

Figure 10. Wind analysis for section 2

IOP Conf. Series: Materials Science and Engineering **401** (2018) 012004 doi:10.1088/1757-899X/401/1/012004

3.5. Energy Performance

Project Summary

Location and Weather	
Latitude	30.09°
Longitude	31.23°
Summer Dry Bulb	40°C
Summer Wet Bulb	19°C
Winter Dry Bulb	8°C
Mean Daily Range	16°C

Building Summary

Inputs	
Building Type	House
Area (m^2)	308
Volume (m^2)	877.95
Calculated Result	
Cooling Load Density (W/m^2)	43.01
Heating Load Density (W/m^2)	105.29

From the building design, it can be predicted that the cooling and heating load per m2 of the building was 43.01W/m2 and 105.29 W/m2. According to Passive House Standard, heating and cooling load of a building should not exceed 15kWh annually or 10W/m2 [12]. This was 10% of energy consumed from conventional building, that means conventional building consume 100W/m2 for heating or cooling.

It can be concluded that the cooling load in the building was lower than conventional building. However, the heating load was relatively the same with the conventional building. The design was still inefficient to reduce heating load, thus another passive cooling strategies should be applied to reduce more energy consumption.

4. Conclusion

In summary, the main factor that should be considered when designing building in hot dry climates is orientation. Orientation determine the design of building elements, such as the length and the width of building shape, the position of opening windows, air ventilation, sun shading, and the slope of roof. Great design of building elements will produce high performance of a building where it has less energy consumption. The best building orientation was determined by sun path and wind direction.

The position of opening window is determined by the sun orientation. Sun shading in the building surface is important. It also determines the heating and cooling load of the building. The more opening windows are applied, the larger heating and cooling is produced. South area was the only unshaded area during summer and winter. Therefore, less opening should be applied in the south façade of the building to reduce the solar heat gains.

Air ventilation should be decided by the direction of the wind. The larger the air ventilation, the more wind comes to the building. Prevailing wind comes from south-west and north-east axis. Thus, the natural ventilation could be applied in this axis to catch more cool air into the building.

The most suitable building shape for hot-dry climate is compact building with U-shape and inner court yard in the middle. This building type create more shading, thus reduce the solar heat gains.

From the result, the application of traditional element was relatively effective as passive design strategies to reduce solar heat gains and applying natural ventilation because it has less opening in the south, catch the breeze air from south-west and north-east axis, while consumed less heating load. However, the cooling load was relatively the same with the conventional building, this might be lower by using another passive design strategies.

5. Acknowledgement

Authors would like to dedicate gratitude to Mohamed Gadi, Department Architecture and Build Environment, The University of Nottingham for assisting during the research process. We gratefully acknowledge the funding from USAID through the SHERA program - Centre for Development of Sustainable Region (CDSR).

6. References

[1] NOAA, "National Oceanic and Atmospheric Administration," NOAA, 2018. [Online]. Available: ftp://ftp.atdd.noaa.gov/pub/GCOS/WMO-Normals/TABLES/REG__I/UB/62366.TXT.. [Accessed 19 January 2018].

[2] W. M. Organization, "World Weather Information Service," Organization, World Meteorological, 2018. [Online]. Available: http://worldweather.wmo.int/en/city.html?cityId=248. [Accessed 19 January 2018].

[3] C. H. C. A. Group, "Cultural Property Training Resource Egypt," CENTCOM Historical/ Cultural Advisory Group, 2018. [Online]. Available: https://www.cemml.colostate.edu/cultural/09476/egypt02-08enl.html. [Accessed 20 January 2018].

[4] N. A. G. Mohamed and W. H. Ali, "Traditional Residential Architecture in Cairo from a Green Architecture Perspective," *IISTE,* vol. 16, pp. 6-26, 2014.

[5] V. Olgyay and A. Olgyay, Design with climate, 1963.

[6] H. Fathy, Vernacular Architecture: Principles and Examples with Reference to Hot Arid Climates, 1986.

[7] A. Salama, "A Typological Perspective: The Impact of Cultural Paradigmatic Shifts on the Evolution of Courtyard House in Cairo," *METU-JFA,* vol. 23, pp. 41-58, 2006.

[8] B. Abouseif, "Note sur la Fonction de la Cour dans la Maison Moyenne du Caire Ottoman; in (eds)," *I'Habitat Traditionnel dans les Pays Musulmans autour de la Mediterrance Cairo, Institut Francais d' Archeologie Orientale,* 1990.

[9] S. Gianni, "Climatic Design in the Arab Courtyard House," *Environmental Design: Journal of the Islamic Environmental Design Research Centre,* vol. 1, pp. 82-91, 1988.

[10] B. S. Hakim, Arabic-Islamic Cities: Building and Planning Principles, London dan New York: Kegan Paul International, 1988.

IOP Conf. Series: Materials Science and Engineering **401** (2018) 012005 doi:10.1088/1757-899X/401/1/012005

Lighting Heritage Building Practice in George Town, Penang Island

[1]Safial Aqbar Zakaria [2]Mohd Jaki Mamat [3]Lim Shu Chern [4]Chong Mun Hong [5]Wang Zi Xuan [6] Siti Nur Ayat Shafie
School of Housing, Building and Planning, Universiti Sains Malaysia, 11800, Penang
E-mail: [1]ssafial@usm.my, [2]mohdjaki@usm.my , [3]winc.line2@gmail.com, [4]840493606@qq.com, [5]shuchern@hotmail.com

Abstract. In this technological era, lighting fixtures are can be seen anywhere in the city to accommodate its specific functions or to achieve some purposes in any forms and colours. The patterns and arrangements of the lighting fixtures on the buildings and structures at night creating a sense of art which emphasizes the nature or the potential of the buildings. Nowadays, lighting designers tend to put more lighting fixtures into their design to give a better outlook and as an attraction for tourism at night. In some manners, light art is like a rose that has thorns on it. The public nowadays appreciate the beauty of light art without realizing the harm that stays under the exposure of colourful bright light designs. However, light art has already become a piece of puzzle that cannot be missing in the growth of economy of developing country in tourism. The objectives of this study are to highlight the rules and regulations that must be obeyed in applying lighting fixtures on historical buildings and to suggest some recommendations to meet the balance between the beauty of light and the appreciation of historical buildings in George Town.

Keywords: Lighting, historical building, urban lighting, lighting pollution, awareness

1. Introduction

Historical buildings in George Town, Penang Island show its importance and uniqueness in every particular place are ~~are~~ being appreciated more by tourists today. Many conservation works and renovation works to add in lighting fixtures has made the buildings become more practical and aesthetic twenty-four seven. Even though the lighting decorations might be an attraction to the tourists but unfortunately some buildings were installed with excess lighting fixtures without realizing it. Excessive lightings will lead to light pollution that might also bring a lot of harms towards human, animals, plants and also the biodiversity of the particular area. Above all of these, does the design of renovation or improvement plans met the guidelines that is fixed by the The United Nations Educational, Scientific and Cultural Organization (UNESCO) on heritage buildings? This paper will further discuss on some effects on lightings at night and also some suggestions towards the lighting fixtures that can balance the function and the aesthetics of the heritage building of George Town, Penang Island.

Content from this work may be used under the terms of the Creative Commons Attribution 3.0 licence. Any further distribution of this work must maintain attribution to the author(s) and the title of the work, journal citation and DOI.
Published under licence by IOP Publishing Ltd

IOP Conf. Series: Materials Science and Engineering **401** (2018) 012005 doi:10.1088/1757-899X/401/1/012005

2. Objectives

1. To identify the good and poor lighting based on the observation and findings for future design references.
2. To highlight the rules and regulations that must be obeyed in applying lighting fixtures on historical buildings
3. To understand types of lighting existed on historical building and their pros and cons.

3. Research Method

Primary data was collected through observation of the building exterior by the researchers at the site, which is the area around Penang Island City Hall to gauge the nature and extent of the building's lighting issue. Observation method is used to study the issue of lighting design at the place. Through observation, researchers review the characteristic of the building and the lighting fixtures that were applied to the building, the arrangement of the lighting fixtures and the functionality of the lighting system as well as its impact to the building and the surrounding. Observation had been done on daytime and at night to determine the difference between the appearance condition of building being lit out under daylighting and artificial lighting. Pictorial documentation entailed capturing in digital format the whole architectural structure and focus on the lighting fixture applied on the exterior of the building. Library research method is used to collect data and references regarding the research. This method was used with both electronic and printed repositories of information. Any resources that are related to the issue and the context of the research topic, for instance reference books, journal articles, internet articles and the information ollected were evaluated and defined to give definition to the research outcomes.
Quantitative method was used to collect data on the illuminance of the lighting fixtures at different angles and distances to the surfaces buildings. The numerical data collected are analysed and used to explain the impact on lighting fixtures to the building and the surrounding. Some statistical analysis study data was used to compare and contrasted with the data collected in order to reiterate the lighting issue being investigated.

4. Description of Heritage Building

The heritage building that is ınvolved in the research is the Penang Municipal Office in Penang Island. The building is the local government headquarters located next to The Esplanade of Penang, which is a waterfront area on the northern coast of George Town. It was the first area where Francis Light landed in Penang in 1786. It is one of the UNESCO World Heritage Site on Lot number 69, Section 19, Georgetown, Northeast District of Penang, Malaysia. The two-stories City wall is a masonry building with interesting timber featuring on its interior and exterior. The building formerly known as Municipal Office, then its name changes into the City Hall of George Town when city status achieved by George Town and its name redefined as Penang Municipal Office after the head office of the Penang Island City Council moved to the building (Tye, 2015). Penang Island City Hall is a Neo-classical style British colonial architecture that constructed in 1903 and opened to public in 1906. The building was listed as a national monument under Antiquities Act 1976 by the Federal Government due to its significant historical and architectural values (Tye, 2015). The building reflects both Edwardian Baroque and Palladian architectural styles. The building is totally whitewashed, grounded opposite to the Padang Kota Lama together with the Penang Town Hall building, they give an elegant feeling to the panoramic view of the Esplanade.

5. Findings and discussion

Social impacts on implementations of lighting fixtures

The implementations of lighting fixtures onto historical buildings has actually lead to contrary effects towards social impacts. Installing lighting fixtures on historical building is beneficial to the society and country by making a new look for the historical buildings and not only functional as an attraction during daytime but also at night. Lighting application also gives a new and unique identity for the building to serve its purpose. This will help in growing the country's economy through tourism. Other than highlighting the beauty of the detail on historical buildings, lighting fixtures should also play another crucial role which is light up surrounding spaces to avoid the number of crime cases. In contrast, lighting fixtures that lack of considerations may lead to a lot of health issues towards humans, animals, and also plants. For example, cancer, disturbance on circadian rhythm and melatonin production, and optical health issues (Scientific Committee on Emerging and Newly Identified Health Risks).

Major function and goal of implementation of lighting systems

For most kind of lighting systems, the luminous effect is determined by 4 factors: the lighting source's intensity, directional and characteristic; the geometry where the light interact with the receiving surface; the receiving surface that modify light to becomes secondary light by reflection, redirection and colouring of light: and the person who views the lighting source and illuminated surfaces as he or she moved around (Santen, 2006). The main function of lighting systems is to create a focus to the human eye as an attraction towards the shape and structure of the building. Detailing and the structure of historical building is the soul to proven the presence of histories and timeline. Lighting plays an important role on enhancing the details of the historical building through the contrary effect of light and shadows to carry out the essence and the value of the presence of historical building.

Characteristic of the visual environment

"Our eyes are constructed to enable us to see forms in light" said Le Corbusier. Light is an architectural material, but an intangible one, building form is not perceptible without light to reveal it. The lighting fixtures that were installed on buildings are intended to emphasize the structural form of the building, however the truth is that sometimes it does and sometimes it does not. Distinguishing form through patterns of light tells public how the parts of a building interprets its characteristics. The details of the building were clearly shown under the uniform ambience of daylight in the day (Figure 1 & 3). The fine carving of the Penang Municipal Office cast shadows onto the exterior surface and bring out the Neo-classical style of a British colonial architecture. Daylight reveals the subtleties of the structural form while shadow appearing to cast upon it.
The ambience of the Penang Municipal Office is uniformly illuminated during nighttime and the installation of lighting fixtures does create a focus on certain details of the buildings (Figure 2 & 4). However, the arrangement of the lighting projected has caused glare from the reflection on the white surface of the columns which affect the details of the historical buildings hard to be distinguished. In fact, in the spectrum of light wavelength, if the color of a surface is anything other than white, it means that it absorbs light of some wavelengths (Deziel, 2017).

Type of light source and design arrangement

The type of lighting fixture that installed on the Penang Municipal office is LED floodlight. The colour of the projected light is white colour light. According to the guidelines given by UNESCO on heritage buildings, floodlights used on building must be natural white light. White light is defined as the complete mixture of all of the wavelengths of the visible spectrum (Cardenas, 2015). According to Cardenas if the beams of light of all of the colours of the rainbow onto a single spot, the combination of all of the colors will result in a beam of white light (2015). White light is in the range of visible light spectrum which is 400 x 10^{-6}mm - 700 x 10^{-6}mm. (nanometer (nm) = 1 x 10^{-9}m)

Table shows the the full spectrum that forms white light (Cardenas, 2015):

Table 1: The light spectrum

Color	Wavelength		Color	Wavelength
Violet	380-450 nm		Yellow	570-590 nm
Blue	450-495 nm		Orange	590-620 nm
Green	495-570 nm		Red	620-750 nm

The design arrangement of the floodlights is located in between of every two columns and also the column of entrance. The placement of the floodlight caused glare that reflects from white color surface of the column. The detail of building facades of Penang Municipal Office was unable to distinguish from certain angle due to over exposure of floodlight.

Key design factors and criterions of lighting systems

It is unnecessary to use high light intensity floodlight to give impact on the details of the historical building facades. Uniformly projected lights such as daylight is more than enough to enhance the beauty of the detail on the building facades by the contrary effect of light and shadows. The light illuminance level for day can meet up to 10,000 lux. However, it might be a different in situation at night. According to the illuminance level which was collected using the lux meter, the illuminance from the spotlight on the field shows an illuminance level of approximately 10 - 40 lux depending to the distance from the source. The ambience of the building was softly illuminated and faded radiantly along the distance. However, the eyes will function differently at night. Illuminance level of 50 lux on the area is already sufficient to distinguish the building and the detail of building facades due to reflective white surface of Penang Municipal Office.

Therefore, there are some design considerations that can be considered when implementing floodlight into designs which are type of light source, reflectivity, color, and texture of the projection surface, distance of the surface from the light source, and color filter used.

For the type of light source, lower power input lighting fixtures can be used to reduce the intensity of light and also more energy saving and energy efficiency. Generally, lamps with higher wattage produce more light. Light output could range from 2600 to 135,000 lumens (1 lux = 1 lumen/m^{2}-). Color temperatures also affect the brilliance. The surface of the wall of Penang Municipal Office is white in color paint and the most reflective surface among other colour surfaces under the same material and textures. White painted wall can have a better appearance under floodlight compared to metallic material surfaces.

IOP Conf. Series: Materials Science and Engineering **401** (2018) 012005 doi:10.1088/1757-899X/401/1/012005

The distance of the surface from the light source affects a lot on the projecting effect of floodlight. The distance of the surface from the light source will increase the illuminance level towards the surface. In this situation where the reflected light on white surface will cause glare that will lead to discomfort for human eyes for long period of exposure. On the other hand, color filters can also be helping in reducing the white light intensity as all color filters (gel and glass) absorb light. The color filters, to some degree, color is traded for brightness. A white light source (at least 3200ºK) must be used, otherwise, the colors will be distorted. Since color light is prohibited in the guidelines, polaroid can be used to reduce 50% of the light intensity towards the surface of the building. This can reduce the glaring effect as well as easier for the eyes to distinguish the detail of the building facades. The public can enjoy and appreciate the beauty of historical building facades and can see it clearly and without discomfort from glaring issue.

Figure 1 & 2-: Penang Municipal Office in Penang Island

Figure 3 & 4-: Penang Municipal Office in Penang Island

IOP Conf. Series: Materials Science and Engineering **401** (2018) 012005 doi:10.1088/1757-899X/401/1/012005

6. Conclusion

Light acquires meaning in architecture relationally. It is related with the transfer of meaning from the initial intention of the designer into the personal expression of the inhabitants. While personal experience for light will be the interpretation for both. The researchers are intended to explore and present light in a cohesive way as it is interconnected closely with architecture and our daily life. Designers must always try to approach light in an integrated way, both conceptually and practically so that light becomes more directly and meaningful. Light is only one of the aspect in architecture realm, but light plays vital and essential roles in revealing a building, its intentions, its structure, its form, its appearance and its meaning. Light always reveals architecture and the best instances, architecture reveals light. The appreciation on the beauty of light must not neglect the balance of the system in the historical area. In this modern and technology edge-leading era, many new solutions will be coming up to overcome the issues that we should be concerned.

References

[1] Deziel, C. (2017). Which colors reflect More Light? Retrieved from https://sciencing.com/colors-reflect-light-8398645.html

[2] Cardenas, R. (2015). White Light: Definition, Source & Spectrum. Retrieved from https://study.com/academy/lesson/white-light-definition-source-spectrum.html

[3] Tye, T. (2015). Penang Island City Hall. Retrieved from http://www.penang traveltips.com/city-hall.htm

[4] Santen, Christa van. (2006). Light Zone City – Light Planning in the Urban Context. Birkhaeuser: Bael.Boston. Berlin, p. 10.

[5] Millet, M. S. (1996). Light Revealing Architecture. Van Nostrand Reinhold: New York.

[6] Neumann, D, Champa, K. S. (2002). Architecture of the Night. Prestel Publishing: Germany.

[7] Scientific Committee on Emerging and Newly Identified Health Risks (SCENIHR). (2012). Retrieved from https://ec.europa.eu/health/scientific_committees/emerging_en

Territorial Functioning and Victimization in Housing Area

Syarmila Hany Haron[1]

School of Housing, Building and Planning, Universiti Sains Malaysia, Penang, Malaysia
Telephone: 013-4353310, Fax No.: 604 657 6523
Email: syarmilahany@usm.my

Abstract: Territorial functioning represents a collection of behaviors, markers, and attitudes expressed by people in small groups as a way to manage their environments and the spaces they physically occupy. This paper examines attitudes and marking behavior associated with territorial functioning by residents of two housing areas in Malaysia that differ in terms of land uses and in residents' perceptions of the level of crime in their neighborhoods. In the study, 144 participants participated in the survey. They reported levels of crime similar to the ones reported in the British survey of crime in 2015. The data generated by the survey were then analyzed using PSS; descriptive and structural equation modeling that combined the three elements of territorial functioning—behavior, markers, and attitudes—to create an overall measure of territorial functioning._The results of the analysis indicated that territorial functioning is negatively associated with the number of break-ins of homes in a neighborhood regardless of the type of neighborhood.

Keywords: Territorial Functioning, Crime Prevention, Victimization, Fear of Crime.

1. Introduction

Crime is a significant concern in every country in the world, and numerous scholars, law enforcement agencies, and policymakers are involved in multiple efforts to reduce crime and enhance safety for citizens. Naturally, a primary concern is the level of police protection provided in terms of the number of officers deployed for resident patrols. Effects of an inadequate ratio of officers to citizens on curbing residential crime have been heavily debated and the issues are still not been resolved [1]. According to Interpol [2], the ideal ratio of officers to citizens is 1:250. In Malaysia, there are approximately 120,000 officers and a population of 31.7 million, resulting in a ratio of 1:270.

Another approach to crime prevention in Malaysia is a "neighborhood watch" style program called "Ops Cantas" that includes various policing strategies and is being implemented nationwide [3]. It was introduced in Malaysia on October 2013 and led to a reduction in break-ins of 456 cases in 1999 and 1,661 cases in 2000 and a decline in burglaries of 914 cases in 1999 and 2,346 cases in 2000 [1]. According to Taylor (1988), there are three elements that are significant in Territorial functioning, which is territorial attitudes, territorial markers and territorial behavior are interwoven and responded to support and provoke one another. If one of the elements should be higher as well. Taylor (1988) clearly noted that

Content from this work may be used under the terms of the Creative Commons Attribution 3.0 licence. Any further distribution of this work must maintain attribution to the author(s) and the title of the work, journal citation and DOI.
Published under licence by IOP Publishing Ltd

territorial functioning was highly variable across different communities, although it was also essential to a community's long-term health (Frost *et. al.,* 2000). As such, occupants who join a Neighbourhood Watch anti-crime group, might define the residential block as a secondary territory that would respond defensively to strangers who would use the block as a public territory.Territorial functioning has been proposed as an effective way to reduce crime in Baltimore. [4]. In his foundational work, Taylor [4] defines territorial functioning as a system of attitudes, sentiments, and behaviors that are specific to a clearly marked location. These behaviors, patterns, and markers signify that a group of residents has some expectations of exclusivity of use of a space, as well as responsibility for and control over activities in that space. Provided the hot weather even the spacious compound does not persuade the residents to remain outside for long. Only half of them do stay out of the house frequently. Half of these residents interacted since remained outdoors had increased the opportunity for interaction and surveillance. Where physical design had been known to affect social activities outside home garden in terms of security, this should be taken into serious consideration in the future planning of houses. Houses should be constructed such a way that there is maximum opportunity for interactions and surveillance. The prevalence of crime at the corner lot homes frequently account provided that it is easy opportunities for criminals to break-in and scurried off after the criminal attempts. Physical appearance with high maintenance of garden, demarcation and territorial displays is also important for the burglars to determine the signals of valuables inside home as prominent and targeted for burglary [4].

2. Materials and Methods

The primary objectives of the study are to: (1) identify associations between residents' demographic characteristics and the outcome of territorial functioning and victimization, (2) examine any differences between two types of neighborhoods in territorial functioning, fear of crime, and perceptions of the severity of crime, and (3) analyze any associations between the types of offenses that occur and types of neighborhood in which they occur. Figure 1 presents a model of the variables analyzed in the study.

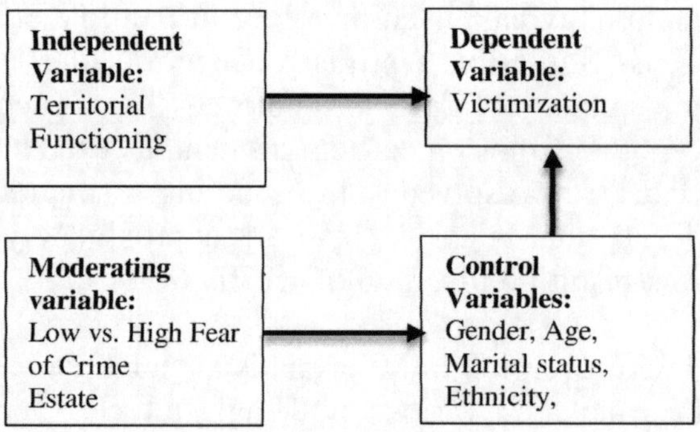

FIGURE 1

Two neighborhoods in Selangor, Malaysia occupied by both low-income and high-income residents were selected as study sites. They are typical of a majority of ungated landed residential properties in urban Malaysia. Both are situated mainly in residential areas with commonly available basic shopping facilities provided within the estates. The survey asked respondents to rate the degree of territorial functioning to which they agreed with three statements describing territorial functioning (knowing their neighbors' names, feeling responsible for their neighbors' properties in their absence, and feeling comfortable living among their neighbors), how the level of fear of being victims of five crimes (burglary, assault, sexual harassment, rape, and vehicle theft), and the degree to which they felt five other crimes occurred in their neighborhood (burglary, car theft, vandalism, drug abuse, and hooliganism) (see Table 1). A cross-tabulation analysis was used to partition the respondents' profiles by type of estate—a high or low fear of crime. The analysis showed that none of the control variables (gender, age, marital status, education, and income) varied significantly between the two estates. This is an important measure in ensuring that the sampling strategy is not biased.

3. Results

Territorial Functioning, Fear of Crime, and Perceived Crime

To identify differences in the two estates in *territorial functioning*, *fear of crime*, and *perceived crime*, a nonparametric Mann-Whitney U-test was conducted using the responses from the survey. The nonparametric test was selected because the study variables were ranked ordinally for the nominal estate type. The results of the analysis are reported in Table 1. The analysis indicates that none of the measures of territorial functioning varied with the type of estate. Likewise, there were no significant differences between the high-crime and the low-crime estates in terms of the five fear factors of being *burgled, assaulted*, *sexually harassed*, and *raped* and of having *a vehicle stolen*. Thus, it appears that the two estates shared a similar degree of fear about those crimes. There were differences, however, in their perceptions of the prevalence of the five types of crime in their neighborhoods. Residents of the high-crime estate perceived the rates of burglary, car theft, vandalism, drug abuse, and hooliganism as significantly higher than residents of the low-crime estate. The comparison confirms that the two neighborhoods selected for study were indeed heterogeneous in terms of some perceptions of crime.

TABLE 1

Territorial Functioning, Fear of Crime, and Perceived Crime in the Low-crime and High-crime Estates

	Low-crime Estate	High-crime Estate	Mann-Whitney U-test
Territorial Functioning			
I know the names of most of my neighbors	95.36	76.35	510
I feel responsible for watching over my neighbor's house when they are on holiday	90.25	91.20	465
I feel comfortable living among my neighbors	89.54	85.14	419
Fear of Being a Victim of a Crime			
Being burgled	288	167	98
Being assaulted	15.22	13.25	91
Having a vehicle stolen	14.66	11.72	74.3
Being sexually harassed	13.79	15.21	83
Being raped	15.78	17.79	91.5

Perception of the Prevalence of a Crime			
Burglary	81.23	53.10	1520**
Car theft	78.42	56.71	1567**
Vandalism	77.76	56.34	1587**
Drug abuse	77.08	61.39	1685**
Hooliganism	76.27	58.28	1726**

** $p < 0.01$, * $p < 0.05$.

Victimization

A second cross-tabulation analysis was conducted to determine if there was any association between being a victim of five types of offenses and the type of estate. Those results are reported in Table 2. The analysis identified that there was no significant difference between the low-crime and the high-crime estates for tampering with vehicles and stolen parts. However, residents of the high-crime estate perceived the number of successful and attempted break-ins as much greater than residents of the low-crime estate.

TABLE 2

Victimization: Low-crime vs. High-crime Estate

		Estate					
Type of Crime		Low-crime (n = 58, 40.28%)		High-crime (n = 86, 59.72%)		Total	
Parts stolen from vehicles	Yes	16	(53.3%)	14	(46.7%)	30	(21.5%)
($\chi^2 = 0.54$)	No	31	(36.9%)	53	(63.1%)	84	(78.5%)
Vehicles tampered with or damaged	Yes	10	(25%)	30	(75%)	40	(35.7%)
($\chi^2 = 2.7$)	No	30	(41.7%)	42	(58.3%)	72	(64.3%)
Burglary[†]	Yes	2	(28.6%)	5	(71.4%)	7	(4.9%)
($\chi^2 = 3.58$)	No	72	(52.6%)	65	(47.4%)	137	(95.1%)
Unsuccessful attempt at burglary*	Yes	4	(19%)	16	(76.2%)	21	(14.6%)
($\chi^2 = 5.022$)	No	61	(49.6%)	62	(50.4%)	123	(85.4%)
Defaced or damaged house*	Yes	5	(33.3%)	10	(66.7%)	15	(13.6%)
($\chi^2 = 4.749$)	No	52	(49.5%)	53	(50.5%)	105	(95.5%)

[†] $p < 0.10$, ** $p < 0.01$, * $p < 0.05$.

Relationship between Territorial Functioning and Crime

In the third analysis, the model described in Figure 1 was run using the data from the survey to determine whether the relationship between territorial functioning and crime was affected by the type of estate. Those results are presented in Table 3 and further illustrated in Figure 2. The results show that territorial functioning are negatively correlated with the number of break-ins in both estates.

TABLE 3

Hierarchical Regression of the Dependent Variable: Number of Trespasses

Variables	Model 1 Territorial Functioning (TF)	Model 2 TF + ET	Model 3 TF + ET + (TF x ET)
Territorial Functioning (TF)	–0.191**	–0.185**	0.167
Estate Type (ET)	—	0.161*	0.745†
TF x ET	—	—	–0.677
ΔF	7.677**	6.007*	1.96
ΔR^2	0.041	0.029	0.008

Notes: Estate 1 = low-crime; Estate 2 = high-crime. ** $p < 0.01$; * $p < 0.05$; † $p < 0.10$. The values for the variables are standardized βs.

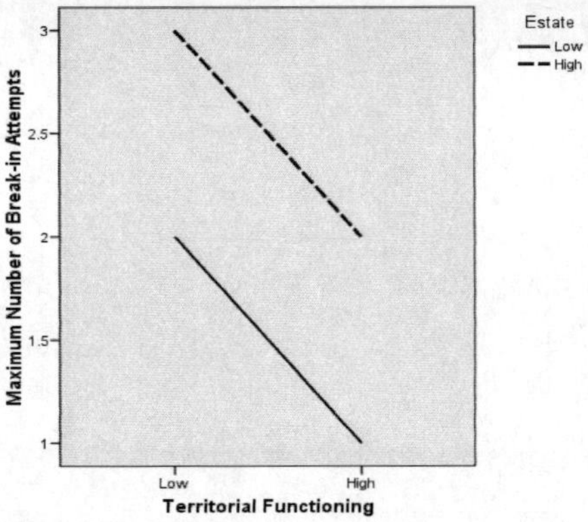

FIGURE 2: The effect of territorial functioning and trespasses: Low-crime estates vs. high-crime estates

4. Discussion and Conclusions

This research is concerned with territorial functioning as a method of crime prevention. Scholars have measured territorial functioning using numerous methods but little attention has been paid to cross-validation of individual features such as socio-economic characteristics. Malaysia is an important area for studies of territorial functioning because territorial attitudes and territorial markers have a stronger influence in Malaysia than in nations in the United Kingdom. The results of this study demonstrate that territorial functioning had a negative impact on incidences of burglary in both low-crime and high-crime estates. Further research on territorial functioning should examine additional settings such as gated versus ungated communities with high and low rates of crime. Future studies could also broaden the analysis of various types of territorial functioning to provide a more comprehensive picture of its effects on crime. Such comparative studies could make significant contributions to our understanding of the effects of territorial functioning and crime prevention.

References

1.Bukit Aman Police Headquarters (2002) Statistik Kes Pecah Rumah dan Curi Mengikut Jenis Bangunan Jan–Dis 1999–2001. Kuala Lumpur. Presence: Teleoperators and Virtual Environments 10(6): 583–599.

2.Frost J.D. and D.R. Saussus. (2000). "Efficient Mitigation of Edge Effects in Nearest Neighbor Analysis." Journal of Testing and Evaluation, 28:3-13.

3.International criminal Police Organization INTERPOL (2017) [FR], https://www.interpol.int.

4.Sinar Online (2014). OPS Cantas Khas PGA banteras jenayah. http://www.sinarharian.com.my/global/ops-cantas-khas-pga-banteras-jenayah-1.274355

5. Taylor RB (1988) Human territorial functioning: An empirical, evolutionary perspective on individual and small group territorial cognitions, behaviors and consequences. Cambridge: Cambridge University Press.

IOP Conf. Series: Materials Science and Engineering **401** (2018) 012007 doi:10.1088/1757-899X/401/1/012007

Relationships of Early British Amenities with the Formation of Small Towns in Perak

Mohd Jaki Mamat[1], Muhammad Firzan Abdul Aziz[2]
Senior Lecturer, School of Housing Building and Planning, Universiti Sains Malaysia, Penang, Malaysia [1, 2]

E-mail: mohdjaki@usm.my [1], nazrif87@gmail.com [2]

Abstract. The British was influential in shaping the early urban and town planning of Malaysia. In this regard, several types of amenities in the likes of police station, post-office, administrative building, religious building, educational institution and many more were built during the mid 19th Century until the post-independence era. This study objectively focuses on identifying early amenities built by the British, understanding their relationships with early towns' socioeconomic activities as well as seeking the factors which had cause the arising needs of such amenities. Data were acquired through the means of seminal documents and historical manuscripts as well as interviews, using case studies of 14 early small towns in Perak. The current study found that the British prioritised on providing security-based amenities for the early towns that they have economic interests on. On the other facet, the British also emphasised on communication and accessibility aspects for the early towns located in distance from the main cities in Perak, which socioeconomic activities were based on agricutural industry.

Keywords; amenities, first British built, early small towns.

1. Introduction

One of the factors in urban planning as affirmed theoretically is on the provision of amenities (Ibrahim 2009). Being considered as an important criteria for any township development, amenities are regarded essential be on the aspects of economic expansion, population increment or physical enhancement (Mitra 2015; MacLachlan, 2015; Garretsen, 2017). Their presence signifies a township's significance (Powe, 2012) and is indicative on the township's density level and future potential (Bantjes, 2011). Provision of public amenities nowadays is focused on creating a better and quality lifestyle- as seen through the creation of public parks, sports and recreational areas, cultural enclaves and events, as well as, shopping and entertainment centres (Perloff, 1966; Kloosterman 2014; Garretsen, 2017).

Studies on early amenities in Malaysia are necessary to promote a better understanding on the significance of the country's early townships. As proven historically, the British's arrival in Malaya (Malaysia) Has caused certain transformations in the local town planning system. This scenario is especially apparent in Perak, which was the very first state in the Malay Peninsula intervened by the British in 1874 (post-signing of the Pangkor Treaty). Hence, scrutinising the remaining amenities built during the British era may lead us to the revelation on the significance of the early townships, besides fostering our understanding on their early socioeconomic activities and scenarios.

This paper focuses on early amenities built by the British in 14 out of 22 small townships available in Perak, selected based on their varying economical backgrounds. Qualitative data were gathered through compiling and analysing old manuscripts and historical documents, which later were validated by interviews conducted with the locals.

Content from this work may be used under the terms of the Creative Commons Attribution 3.0 licence. Any further distribution of this work must maintain attribution to the author(s) and the title of the work, journal citation and DOI.
Published under licence by IOP Publishing Ltd

IOP Conf. Series: Materials Science and Engineering **401** (2018) 012007 doi:10.1088/1757-899X/401/1/012007

2. Literature Review

Development of small townships in the past were catalysed by the expansion of early settlements (Wang, 2010). These early settlements were formerly based on agricultural industry (Choguill, 1989) and later transformed into local economic centres (Powe, 2012). Implying from that, public amenities were then equipped by former authoritities to such places to cater their socioeconomic activities and needs (Choguill, 1989). Apart from supporting the agricultural industry, public amenities were also built to cater community needs through the provision of post office, day market, assembly hall etc. (Mayer, 2010). In turn, such places inreased in significance and became the centre of attention for both locals and outsiders (Yamu, 2015).

Prior to the British era, the royal authorities were the influential parties in shaping the early townships and traditional settlements. During this period, they were in direct control of constructing public buildings which mostly concerned with the aspects of defence and security of the royal palaces. Mosque buildings were highly significant as a public amenity at that time. Apart from their primary function as a place to perform Islamic congregational prayers, mosques were multi-functioning as educational centre as well as for ceremonial and gathering place (Harun, 2012). Only after the British's arrival, public amenities were provided to commensurate their interests which further cater the local needs (Sanusi, 2009; Taylor, 2010; Harun, 2012). It can be understood that the British remained their urban development approach practised in England at Malaya, whereby provision of public amenities at a place was mainly to cater its economic activities (Powe, 2012; Harun, 2012).

Accordingly, Mohd Hairy (2014) theorised that increment of population, due to concentration of settlements into local economic centres, was also influential in the provision of public amenities by the British. The aspects of security, health and recreation were prioritised by the British during the initial stage of town or city formation to ensure the safety of its own citizen as well as other Europeans. Apart from that, the British appllied the theory of racial segregation by building separating wall, gate and security hut for the sake of safety (Home, 1997; Ambe, 2008). These, then were extended to cater healthcare needs through building hospitals for areas affected with diseases such as Malaria for instance (Mizushima, 1997; Ambe, 2008).

It is evident that economic factor was influential in the formation of early towns by the British in Malaya (Lockard, 1974; Suriati, 2013; William, 2014). Types of amenities provided at early towns were basically dependent on economic activities held at that time (Ambe, 2008; William, 2014). Stability and security aspects then were considered to ensure the smoothness of economic agenda, as apparent in the planning of early settlements whereby racial segregation concept was enabled trough the setting-up of gated community with security huts (Sanusi, 2009; Yuen, 2011). Upgrading of transportation roads and communication system, which is considered as the next phase of town formation (William, 2014), then follow suit as the British intended to enhance inter-towns accessibility and reachability.

Implying from the literature, the two main factors influencing the British to build public infrastructures and amenities are safety and communication, which vary respectively to correspond with the socioeconomic needs of a town or city. Table 1 details out such relationship for the 14 small towns in Perak.

Table 1*;* Early Amenities Built by the British in Small Towns of Perak.

No.	Town/Year	Early British amenities	Sosioeconomic activities
1	Gopeng (1870s)	Police Station (1879) Court House (1880s) Hospital (1884) Post Office (1891) Sanitary Building Office (1890s)	Tin mining
2	Papan (1880s)	Police Station (1887) Post Office (1895) Railway Station (1908)	Tin mining
3	Tanjung Rambutan (1893)	Railway Station (1896) Police Station (1896) Post Office (1896) Court House (1897) Psyciatric Hospital (1911)	Tin mining
4	Pusing (1890s)	Market (1897) Police Station(1890s) Railway Station (1908) Post Office (1924)	Tin mining
5	Sungkai (1900)	Railway Station (1894) Post Office (1905) Police Station (1936)	Agriculture
6	Kuala Kurau (1900s)	Post Office (1908) Fishery Department (1922) Police Station (1920s) Fire Fighting & Rescue (1920s) Jetty (1934) Customs Office (1934)	Fishery & agriculture
7	Padang Rengas (1900s)	Post Office (1894) Police Station (1937) Railway Station (1950s)	Agriculture
8	Karai (1900s)	Post Office (1898) Police Station (1905)	Coal mining
9	Klian Intan (1909)	Police Station (1909) Post Office (1910) Hospital (1910s)	Tin mining
10	Beruas (1910)	Post Office (1913) Police Station (1920s) Court House (1930s)	Agriculture
11	Lenggong (1915)	Post Office (1904) Police Station (1930s) Public Work (late 1930s)	Agriculture
12	Lumut (1920s)	Jetty/Port (1920) Post Office (1933) Police Station (1930s) Hospital (1930s)	Trading, fishery & agriculture
13	Parit (1928)	Post Office (1911) Public Work (1920s) Police Station (1940)	Agriculture
14	Bagan Datuk (1931)	Police Station (1890) Jetty/port (1890s) Post Office (1908)	Trading, fishery, agriculture

3. Analysis and Findings

Interpretively from Table 1, the built years of the first shop houses row can be highly associated with the existential years of their respective towns. Although it can be understood that those towns had emerged from the development of traditional settlements, it is imperative to note that shop houses, nevertheless, were either built in separate areas or close to the original settlements. As a result, the first row of shop houses had transformed to a new place of central attention and later developed into townships which remains until today.

Based on the trend yielded, the British apparently favoured in providing amenities in the forms of police station and post office to the small towns. Although school buildings existed at that time, they were however built by the locals and Chinese immigrants in order to carry their vernacular educational system respectively.

Meanwhile, religious buildings (churches) were initially built for the Europeans and not to cater for the locals. Other socioeconomic-based amenities such as market, hospital and shipping port were not being considered as these amenities merely involve the townships partially. They were not significant in relative to police station and post office which functionally cater the entire townships. Police stations were crucial for the sake of stability and protection of the small towns, especially when military bases at that time were only placed at main cities of Perak. Post offices meanwhile were such an important and effective communicational needs following the past limitations in public transportation and traffic system.

The British basically prioritised on stability and security aspects for towns that they had economic interests on. This scenario is apparent at Gopeng, Pusing, Papan and Tanjung Rambutan which were the chains of Lembah Kinta's mining spot, formerly renowned as the world's largest tin producer during its peak era in the late 19th and 20th Centuries. The feud between the Siamese (Thailand) and the State of Perak in acquiring Klian Intan, known as the largest tin mining spot at Perak's northern side meanwhile has led in the intervention of the British. As a resolution, the Bangkok Treaty 1909 was met by both of the British and Siamese, causing the latter to release Hulu Perak District back to the State of Perak. Subsequently, police station was built as a security measure to ensure the place's stability.

Papan Police Station (1887).
Source; National Archives of Malaysia

Lenggong Police Station (1930s).
Source; Lenggong Police Station.

On the other facet, Bagan Datuk was significant as the largest coconut producer in South East Asia. There were 13 large-scale farms owned by the British and other Europeans over there, which products were marketed to all over Europe and the United States. Security aspect was also emphasised in Lumut which is a part of the Straits Settlements comprising Penang, Malacca and Singapore. Before Lumut was returned to the State of Perak in 1934, the town became an international trading port despite its smaller size in relative to other places of the Straits Settlements. Conclusively, the British prioritised on the aspects of political stability and security for the towns that benefiting them economically.

The British also emphasised on communication and reachability aspects for emerging towns in Perak, particularly to providing post office. In total, seven towns were first equipped with post office instead of other types of amenities namely Sungkai, Kuala Kurau, Padang Rengas, Karai, Beruas, Lenggong and Parit. These agriculural and fishery-based towns basically share a common similarity in terms of their remote locations, in which they were situated away from the main cities of Perak such as Ipoh, Kuala Kangsar, Teluk Intan and Taiping for instances. In this regard, communication system was emphasised to ensure the smoothness of British administration at these places. In addition, railway network provided across Karai has enabled the British to manage and supervise the coal industry activities at the town from a more developed city which is in close proximity namely Kuala Kangsar.

The current study also found that police station and post office were built on the very same year for three towns namely Tanjung Rambutan, Klian Intan dan Lumut. The underlying reason for such case can be associated with the construction of railway station in Tanjung Rambutan which necessitated an efficient communication system. In this manner, Tanjung Rambutan can be linked appropriately via railway transportation with Butterworth (Penang) and Singapore. Simulataneous development of Klian Intan and Lumut meanwhile can be linked with the political events occured at those towns, based on the Bangkok Treaty 1909 and the Pangkor Treaty 1874 respectively.

Existing conditions of the abandoned Bagan Datuk Post Office (left) and Tanjung Rambutan Post Office (right).

4. Conclusion

In a nutshell, a town's significance in the past has a direct relationship with its economic contributions. The British administration had prioritise mainly on stability and security aspects to protect their interests, particularly for small towns of mining-industry and port-based. The former aspect was essential for the British in ensuring those small towns to remain economically contributive and stable. The latter aspect meanwhile was inevitable mainly for political needs and reasoning, especially when the British and other Europeans were trading businesses during the formation of those early towns. Communication aspect meanwhile was prioritised by the British to ease their administratitive agendas at small towns in distance from the main cities. Since these small towns were mainly agricultural-based, communication aspect was prioritised instead of stability and security aspects considering that they were less profitable in relative to the small towns of mining-industry and port-based. The existence of post offices per se has also enabled the British to be vigilant on the updates and development of those remote areas.

Acknowledgements

The authors would like to appreciate the Federal Department of Town and Country Planning Peninsular Malaysia (Perak) for funding the current studies on morphology of small town in Perak under the project grant received in 2015 and 2016. Special credits also go to the lecturers and students of Ungku Omar Polytechnic, Perak Heritage Society (PHS), local councils and district offices, local communities, respective building owners and individuals who have directly and indirectly contributed in the process and conduct of this study.

References

[1] Ambe J. Njoh 2008 *Colonial Philosophies, Urban Space, and Racial Segregation in British and French Colonial Africa*, Journal of Black Studies, Volume 38 Number 4.

[2] Arup M, Rajnish K 2015 *New Patterns in Indian Urbanization: Emergence of Census Towns*, Environment and Urbanization Asia, 6(1) 18–27.

[3] Belinda Y 2011 *Urban planning in Southeast Asia: Perspective from Singapore*. TPR, 82 (2). Charles L. Choguill 1989 *Small town and development: A tale from two countries*, Urban Studies, 26, 267-274.

[4] Claudia Y & Pierre F 2015 *Spatial accessibility to amenities, natural areas and urban green spaces: Using a multiscale, multifractal simulation model for managing urban sprawl*, and Planning B: Planning and Design, volume 42, pages 1054 – 1078

[5] Craig A. L 1974 *The Southeast Asian Town In Historical Perspective: A Social History Of Kuching, Malaysia, 1820-1970,* The University of Wisconsin, Ph.D Thesis, 1973

[6] Harun, NZ and Jalil, RAJ 2012 *The Morphological History of the Malaysian Urban Form*, IPEDR, vol 48. 24.

[7] Harry G & Gerard M 2017 *Amenities and the attraction of Dutch cities*, Regional Studies, Vol. 51, NO. 5, 724–736.

[8] Harvey S. Perloff 1966 *New Towns in Town*, Journal of the American Institute of Planners, 32:3, 155-161.

[9] Heike M & Paul K 2010 *Small-Town Sustainability: Prospects in the Second Modernity*, European Planning Studies, 18:10, 1545-1565.

[10] Ian MacLachlan and Julia H 2015 *New Town in the Bush: Planning Knowledge Transfer and the Design of Kwinana, Western Australia*, Journal of Planning History, Vol. 14(2) 112-134.

[11] Ibrahim Mohd, Faizah Ahmad and, Wan Norazriyati Wan Abd Aziz 2009 *Exploiting town planning factors in land development; Case study of urban housing in Kuala Lumpur, Malaysia*, , Journal of Facilities Management, Vol. 7 No. 4, pp. 307-318.

[12] Lan W, Ratoola K and Xiangming C 2010 *Building for what and whom? New Town Development as Planned Suburbanization In China and India*, Suburbanization in Global Society, Volume 10, 319–345.

[13] Mohd Hairy Ibrahim, Azmi Ahmad Baharom, Khairi Arifin, Baharuddin Jabar, Mazdi Marzuki & Kamal Kamaudin 2014, *Investigation of the landscape change in old Malaysian railway; Special references to Port Weld-Taiping Railway*, The Social Sciences, 9 (5): 357-361.

[14] Neil A P & Dave B 2012, *Localising activity within the context of relational complexity: exploring the relevance of rural centre and town centre conceptions, Environment and Planning B*: Planning and Design, volume 39, pages 501 – 517

[15] Nigel T 2010 *What is this thing called spatial planning? An analysis of the British government's view,* TPR, 81 (2).

[16] Peter J. L & John P 2008 *Reconstruction planning and the small town in early post-war Britain*, Planning Perspectives, 23:3, 291-321.

[17] Rod B 2011 *"Rural sustainability and the built environment"*, Journal of Enterprising Communities: People and Places in the Global Economy, Vol. 5, Issue: 2, pp.158-178.

[18] Robert C. K 2014 *Cultural Amenities: Large and Small, Mainstream and Niche-A Conceptual Framework for Cultural Planning in an Age of Austerity*, European Planning Studies, Vol. 22, No. 12, 2510–2525.

[19] Robert H, E & FN Spon 1997 *Of Planting and Planning: The Making of British Colonial Cities*, an Imprint of Chapman & Hall, London, ISBN 0-203-46106-1 (Adobe e-reader format), (1st edition)

[20] Suriati A, David J 2013, *Investigating the Mining Heritage Significance for Kinta District, the Industrial Heritage Legacy of Malaysia*, Procedia - Social and Behavioral Sciences, 105, 445 – 457

[21] William W L H, David Yoon Kin Tong and Gerald Guan Gan Goh 2014 *Revitalisation of the Old Township of Ipoh, Malaysia*, Journal of Place Management and Development, Vol. 7 No. 1, pp. 57-73.

[22] Tsukasa M 1997, *A Historical Study on Land Transaction in a Local Town in Malaysia Kuala Kangsar Shop Lots between 1885 and 1995*, Reprin ted from *Regional Views,* Number 10.

NEW DERIVATION METHOD OF REGION SPECIFIC CURVE NUMBER FOR URBAN RUNOFF PREDICTION AT MELANA WATERSHED IN JOHOR, MALAYSIA

W J Tan[1], L Ling[1*], Z Yusop[2] and Y F Huang[1]

[1]Centre of Disaster Risk Reduction, Lee Kong Chian Faculty of Engineering and Science, Civil Engineering Department of Universiti Tunku Abdul Rahman, Jalan Sungai Long, Bandar Sungai Long, 43000 Cheras, Malaysia

[2]Centre for Environmental Sustainability and Water Security, Faculty of Civil Engineering Department, Universiti Teknologi Malaysia. 81310, Skudai, Johor, Malaysia

[*]Corresponding Author: linglloyd@utar.edu.my

Abstract. The Soil Conservation Services (SCS) rainfall-runoff model has been incorporated into many types of software and adopted by Malaysian government agencies for design use. However, hydrologists argue the accuracy of the predicted runoff results from this model in recent decades. Malaysia does not have a regional specific curve numbers available for the use in rainfall-runoff modelling, and therefore SCS-CN practitioner has no option but to adopt its guideline and handbook values which are specific to US region. Unlike the conventional SCS-CN technique, the proposed calibration methodology in this paper discarded the use of CN as input to the SCS model, rearranged the model equation and derived statistically significant CN value(s) of a specific region through rainfall-runoff events directly through the use of inferential statistics. The SCS rainfall-runoff framework can now be calibrated quickly to address urban runoff prediction challenge under rapid land use and land cover or climate changes with the proposed methodology. Between July and October of 2004, the derived λ was 0.0005 while $\lambda = 0.20$ was found to be statistically insignificant at alpha = 0.01 level at Melana watershed in Johor, Malaysia. Optimum CN of 90.45 was derived to represent the rainfall-runoff characteristic of Melana watershed. Runoff prediction error was reduced by 90% and achieved the Nash-Sutcliffe (NS) index of 0.86. The new method can provide CN adjustment guidelines for SCS practitioners and any software which incorporated the model to predict urban runoff.

Keywords: *Bootstrapping, Curve Number, Runoff prediction, Inferential Statistics*

Content from this work may be used under the terms of the Creative Commons Attribution 3.0 licence. Any further distribution of this work must maintain attribution to the author(s) and the title of the work, journal citation and DOI.
Published under licence by IOP Publishing Ltd

1. Introduction

The United States Department of Agriculture (USDA), Soil Conservation Services (SCS) agency developed the curve number (CN) system in 1954. It's Technical Release 55 (TR-55) classified site conditions with CN. However, researchers reported inconsistent runoff prediction results by using CN and SCS runoff model throughout the world in recent decades [3, 11, 22, 23]. Hawkins (1984) [6] reported that forested watersheds have the highest risk in CN misclassification while inappropriate CN selection was identified as the main cause to produce inconsistent runoff estimates. According to Soulis and Valiantzas (2013) [19], rainfall-runoff (*P-Q*) dataset should be the base to derive CN. Several researches even reported that CN was a random variable between storm events [9, 20] while watershed with several soil groups could not be represented by a lumped CN value but SCS practitioners seldom attempt to calibrate the model and subjectively selected CN value in their studies. [15].

2. Study Site and Methodology

Instead of selecting CN, this study used a new approach to calibrate SCS runoff predictive model through the use of inferential statistics and P-Q dataset to derive CN for the study site. SCS model was introduced in 1954 as:

$$Q = \frac{(P - I_a)^2}{P - I_a + S} \tag{1}$$

where

Q = Runoff depth (mm)
P = Rainfall depth (mm)
I_a = The initial abstraction amount (mm)
S = Watershed potential water retention (mm)

The initial abstraction is the rainfall retention prior to beginning of runoff process. SCS hypothesized that $I_a = \lambda S$ where $\lambda = 0.20$. Substitution of $I_a = 0.20S$ simplifies Equation (1) into a form listed below:

$$Q = \frac{(P - 0.2S)^2}{P + 0.8S} \tag{2}$$

where $P > 0.2S$ or else $Q = 0$. Worldwide study results in recent decades showed that Equation (2) failed to produce reliable results while $\lambda = 0.20$ was also under scrutiny [3, 11, 22, 23]. This study adopted the rainfall-runoff dataset from Chan (2005) [1] which was carried out in Melana watershed to demonstrate the SCS model calibration and CN derivation technique. Inferential statistics was used to derive two key parameters (λ and S value) from Equation (1) according to the given *P-Q* dataset. The watershed is located in Johor, Malaysia. The watershed locates between 1° 30' N to 1° 35' N and 103° 35' E to 103° 39' E (Figure 1). Melana River flows from Gunung Pulai in the northern part through the watershed which covers an area of 21.12 km^2. It was projected that more than 60% of the area would be populated to become residential area by 2010 [14].

This study used twenty-seven P-Q dataset collected between July and October of 2004 in order to limit the effect due to rapid land use and land cover change. Bootstrapping BCa procedure was selected for its robustness nature while the inferential ability of its confidence interval can used for the assessments of the 1954 SCS proposal through the Null hypothesis stated below[2, 21]:

Null Hypothesis 1 (H_{01}): Equation (2) is valid at the Melana watershed where $\lambda = 0.20$.

H_{01} must be accepted in order for Equation (2) to be used to model runoff conditions at Melana watershed. Rejection of H_{01} make way to derive new regional λ value because the initial SCS proposal of $\lambda = 0.2$ fails. The *P-Q* data pairs were used to derive S and λ values from Equation (1). The effective rainfall depth (P_e) is the difference between the rainfall depth (P) and initial abstraction (I_a) thus $P - I_a = P_e$ [7-8]. Substitute P_e into Equation (1), it can be re-arranged to calculate corresponding S and λ values for respective *P-Q* data pair. Non-parametric Bootstrapping technique, BCa procedure with 2,000 random sampling [4] was conducted to produce the 99% confidence interval (CI) of the derived λ and S pairs [16]. Supervised numerical optimisation technique was used to select the best λ and S value within the 99% BCa CI [5] in order formulate a new calibrated SCS runoff prediction model for Melana watershed. The runoff prediction accuracy of the new calibrated SCS runoff prediction model will be assessed and benchmarked against Equation (2) in order to quantify the runoff prediction improvement due to the model calibration and CN derivation.

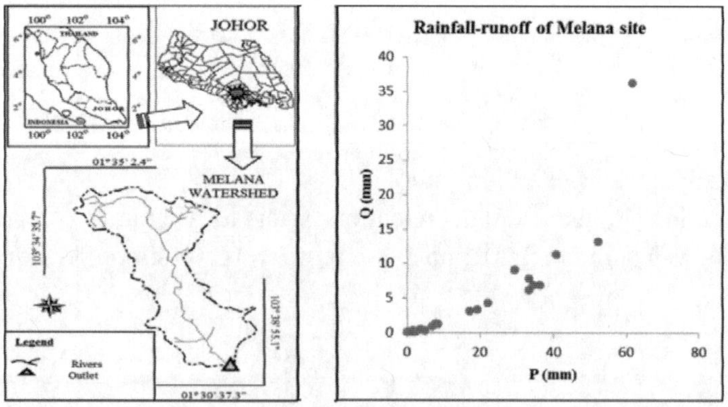

Figure 1. Melana watershed [1] and its P-Q graph.

Equation (1) can be rearranged to solve for $S = f(P, Q, \lambda)$. Substitute $I_a = \lambda S$ into Equation (1) to solve for S will produce the S general formula. If new derived λ value is not equal to 0.20, S general formula can be used to calculate the S_λ value which is different from $S_{0.2}$ (where $\lambda = 0.2$). Any other λ value will result in S_λ which cannot be used to calculate CN. A correlation must be identified between

S_λ and $S_{0.2}$ in order to derive the conventional $CN_{0.2}$ value with the CN formula ($CN = \frac{25{,}400}{S+254}$) which was proposed by SCS [7-8, 10]. The general S_λ formula which we solved is listed below:

$$S_\lambda = \frac{\left[P-\frac{(\lambda-1)Q}{2\lambda}\right]-\sqrt{PQ-P^2+\left[P-\frac{(\lambda-1)Q}{2\lambda}\right]^2}}{\lambda} \tag{3}$$

With the given *P-Q* data pairs and $\lambda = 0.2$, corresponding series of $S_{0.2}$ values were calculated with Equation (3). Bootstrapping (BCa) procedure was used again to generate a 99% CI of the $S_{0.2}$ values while the supervised numerical optimisation technique identified the optimum $S_{0.2}$ value as the best collective representation for the entire school of $S_{0.2}$. The substitution of the optimum $S_{0.2}$ value into $CN = \frac{25{,}400}{S+254}$ produced the SCS $CN_{0.2}$ to represent Melana watershed.

2.1. *Runoff Models Assessment*

Nash-Sutcliffe model's prediction efficiency (*NS*), model residual sum of squares (*RSS*) and model *BIAS* were calculated in order to draw comparison between the calibrated and SCS runoff models with following formulas:

$$NS = 1 - \frac{RSS}{\sum_{i=1}^{n}\left(Q_{predicted} - Q_{mean}\right)^2} \tag{4}$$

$$RSS = \sum_{i=1}^{n}\left(Q_{predicted} - Q_{observed}\right)^2 \tag{5}$$

$$BIAS = \frac{\sum_{i=1}^{n}\left(Q_{predicted} - Q_{observed}\right)}{n} \tag{6}$$

where
n = Total number of data pairs

RSS value indicates the model prediction error. Nash-Sutcliffe value (*NS*) ranges from minus 1.0 to 1.0 while value of 1.0 indicates a perfect model. In the event when $NS < 0$, using the average value of the observed data can predict the dataset better than the model. Zero *BIAS* value indicates a perfect model prediction with no error while the positive value indicates the runoff over-prediction tendency of a model and vice versa.

3. Results and Discussion

3.1. *Null Hypothesis Assessment*

The descriptive statistics of all derived λ values was tabulated in Table 1. Due to the skewed λ dataset nature, the supervised λ optimization study was conducted within the median confidence interval [0.0002, 0.0005] (Table 1). The best λ value was identified to be 0.0005 to formulate the optimum rainfall-runoff predictive model according to the given *P-Q* dataset from Melana watershed.

Table 1. 99% CI Bootstrapping BCa results of all derived λ values

Melana dataset	Descriptive Statistics	(λ) 99% BCa Lower	Upper	Descriptive Statistics	(S) 99% BCa Lower	Upper

	λ			S		
Mean	0.0342	0.0004	0.1331	63.0635	41.8377	86.2105
Median	0.0005	0.0002	0.0005	43.2419	34.5142	93.9196
Skewness	5.1930			0.6960		
Kurtosis	26.979			-0.5450		
Std. Deviation	0.1713	0.0002	0.3222	47.5479	35.4036	93.9196

The skewness of *S* dataset can be considered as normally distributed because its value is near to zero and therefore, the *S* optimisation search was conducted within its mean confidence interval from 41.8377 to 86.2105 mm. The best collective *S* value was identified as 86.2105 mm. As $I_a = \lambda S$ the optimum λ and *S* value yields $I_a = 0.0431$ mm. The calibrated SCS rainfall-runoff prediction model for Melana watershed was formulated as:

$$Q_{0.0005} = \frac{(P - 0.0431)^2}{P + 86.167} \quad (7)$$

where

$Q_{0.0005}$ = Runoff depth (mm) of $\lambda = 0.0005$

As Equation (7) was formed with the optimum λ and *S* value, it will have the same significant level at alpha = 0.01. Not only the standard deviation of λ dataset is not equal to zero but also with high fluctuation percentages hence, λ cannot be a constant as proposed by SCS. None of the λ BCa CI span includes value of 0.2 (Table 1) thus λ cannot be equal to 0.2 for this site. H_{01} can be rejected at alpha = 0.01 level leading to the conclusion that Equation (2) is statistically insignificant and invalid to model runoff at this site.

3.2. The Correlation between $S_{0.0005}$ and $S_{0.2}$ for Melana Watershed

$S_{0.0005}$ and $S_{0.2}$ can be calculated for the *P-Q* dataset using equation (3) through the substitution of respective $\lambda = 0.0005$ and 0.20 corresponding to the same *P-Q* dataset for Melana watershed. Equation (8) has an adjusted R^2 of 0.924, low standard error of 0.457 and $p < 0.001$. Through the substitution of the $S_{0.2}$ parameter in SCS-CN formula ($CN = \frac{25{,}400}{S_{0.2}+254}$), the best representative CN value of 90.45 was derived to represent the given rainfall-runoff condition within 99% CN CI range from 90.45 to 95.12. A statistical significant correlation between $S_{0.0005}$ and $S_{0.2}$ was identified as:

$$S_{0.2} = 0.314 S_{0.0005}^{\ 0.998} \quad (8)$$

where

$S_{0.015}$ = Total abstraction amount (mm) of $\lambda = 0.015$
$S_{0.2}$ = Total abstraction amount (mm) when $\lambda = 0.2$

As shown in Figure 3, Optimum CN of 90.45 formulated a runoff predictive model with lowest *RSS* and the highest *NS* value at $\alpha = 0.01$ level. When $\lambda = 0.2$, the best collective $S_{0.2} = 14.855$ mm for SCS model. The calculated SCS CN value = 94.47. The substitution of $S_{0.2}$ value into Equation (2) formulated the conventional SCS runoff model.

3.3. Runoff Predictive Models Comparison

Runoff models' prediction results were compared in Table 2. Equation (7) has higher *NS* value and lower *RSS* compared to Equation (2). Both model's residual were skewed thus the median residual value was used to assess predictive model's accuracy. Equation (7) has the model residual value nearest to zero and the smaller 99% BCa confidence interval range to indicate that the model is capable of predicting near to perfect (zero residual) runoff prediction.

In contrast, Equation (2) has the tendency to over predict runoff amount (median residual confidence interval range spans within positive figures only). Equation (7) also has lower standard deviation and variance in its model's residual. It also has smaller residual confidence interval range and therefore, it is more stable and reliable while Non-parametric correlation statistics (Kendall's tau b and Spearman's rho) between the predictive model's runoff prediction results and the observed dataset of Equation (7) are also larger than Equation (2) at alpha = 0.01 level to infer that Equation (7) is capable to produce better runoff prediction results than Equation (2).

Table 2. Descriptive statistics comparison of two runoff predictive models

Predictive model's	Equation (7)	SCS: Equation (2)
NS	0.86	-0.36
RSS	198.15	1970.12
BIAS	0.19	4.93
$CN_{0.2}$	90.45	94.47
λ	0.0005	0.20
Residual Skewness	-1.73	1.29
Residual Kurtosis	9.52	0.76
Median Residual	-0.05	0.47
Median Residual: 99% BCa CI	[-0.231, -0.008]	[-0.060, 7.83]
Model Residual Standard Deviation	2.75	7.11
Residual Variance	7.59	50.56
Correlation to Observed dataset:		
Kendall's tau b	0.897**	0.622**
Spearman's rho	0.974**	0.811**

**** Correlation is significant at the 0.01 level (2-tailed)**

On average, Equation (2) over-predicted runoff depth from 7 mm in this study and further magnified toward higher rainfall depths.

4. Conclusions

This study agreed with previous research conclusion that the SCS runoff prediction model can be calibrated with regional specific P-Q characteristics. Statistical assessment rejected H_{01} with 99% confident level. As such, Equation (2) became invalid to model runoff conditions in this study while adoption of Equation (2) will commit a type II error.

This study presented a new CN derivation approach with supervised numerical optimization technique through the guide of inferential statistics. The new calibrated runoff predictive model out-performed SCS runoff model. Residual sum of square (RSS) was reduced by 90%. In comparison to SCS runoff predictive model, it also has higher model efficiency (*NS*), lower *RSS* and *BIAS* to produce the smallest runoff prediction error at alpha = 0.01 level. CN value of 90.45 was derived to represent the P-Q conditions at Melana watershed (between July and October of 2004) with the inherent statistical significance at alpha = 0.01 level.

Direct CN derivation with rainfall-runoff conditions is a swift and economical solution to calibrate SCS runoff model to reflect the latest P-Q condition of any watershed. In the event of rapid urbanisation and climate change, SCS practitioners are able to conduct regional specific calibration for SCS model as proposed to derive region and time period specific CN values for Malaysian watershed.

5. References

[1] Chan C H 2005 *Modelling Watershed Cumulative Effects On Runoff Based On Major Land Uses*

[2] Cox D R 2006 *Principles of Statistical Inference* Cambridge University Press

[3] D'Asaro F, Grillone G and Hawkins R H 2014 *J. Hydrol. Eng.* **19** 04014035 (1–13)

[4] Efron B 2010 *Large-Scale Inference: Empirical Bayes Methods for Estimation, Testing, and Prediction.* (Institute of Mathematical Statistics Monographs/Cambridge University Press) ISBN 9780521192491

[5] Fattorini H O 1999 *Infinite Dimensional Optimization and Control Theory* (Cambridge Univ. Press 1999)

[6] Hawkins R H 1984 *A comparison of predicted and observed runoff curve numbers* In: Proceeding of ASCE Irrigation and Drainage Division special conference Flagstaff AZ: ASCE pp 702–709.

[7] Hawkins R H, Ward T, Woodward D E and Van Mullem J 2009 *Curve Number Hydrology: State of the Practice.* Reston, Virginia, ASCE

[8] Hawkins R H 2014 e-mail communication

[9] Hjelmfelt A T 1991 *J. Hydraul. Eng.* **117** 725-737

[10] Jiang R 2001 *Investigation of Runoff Curve Number, Initial Abstraction Ratio* (University of Arizona)

[11] Ling L and Yusop Z 2013 *A Micro Focus with Macro Impact: Exploration of Initial Abstraction Coefficient Ratio (λ) in Soil Conservation Curve Number (CN) Methodology* IOP Conf.Series: Earth and Environmental Science **18** p 012121

[12] Ling L and Yusop Z 2014 *Inferential Statistics of Claim Assessment* AIP Conference Proceedings **1635** p 805

[13] Ling L and Yusop Z 2014b *Inferential Statistics Modelling and Claim Re-assessment* ICCEMS Conference Proceedings pp 835-840

[14] MPJBT 2001 *Rancangan Tempatan Skudai (1993-2010)* Johor Bahru, Malaysia Pihak Berkuasa Perancangan Tempatan Daerah Johor Bahru.

[15] Ponce V M and Hawkins R H 1996 *J. Hydrol.Eng*. **1** 11-19

[16] Rochoxicz John A Jr 2011 *eJSiE* **4**

[17] USDA-SCS 1964 *National engineering handbook Sect. 4 Hydrology* Washington DC

[18] USDA-SCS 1972 *National engineering handbook Sect. 4 Hydrology* Washington DC

[19] Soulis K X and Valiantzas J D 2013 *Water Resour.Manage*. **27** 1737-49

[20] Van Mullem J A, Woodward D E, Hawkins R H and Hjelmfelt A T 2012 *Runoff Curve Number Method: Beyond the Handbook* U.S. Geological Survey Advisory Committee on Water Information - Second Federal Interagency Hydrologic Modeling Conference Las Vegas, Nevada

[21] Young G A and Smith R L 2005 *Essentials of Statistical Inference* ed Gill R, Ripley B D, Ross S, Silverman B W and Stein M Cambridge University Press

[22] Yuan Y, Nie J, McCutcheon S C, Taguas E V 2014 *Hydrol. Process.* **28** 774-783

[23] Zhou S M, Lei T W 2011 *Calibration of SCS-CN Initial Abstraction Ratio of a Typical Small Watershed in the Loess Hilly-Gully Region* China Agric. Sci. **44** pp 4240-47

IOP Conf. Series: Materials Science and Engineering **401** (2018) 012009 doi:10.1088/1757-899X/401/1/012009

Subjective assessment of thermal comfort for residents in naturally ventilated residential building in Malaysia

Zeyad Amin Al-Absi[1] and Noor Faisal Abas[2]

School of Housing, Building and Planning, Universiti Sains Malaysia 11800, Penang, Malaysia

E-mail: [1] zeyadaminalabsi@yahoo.com, [2] nfaisal@usm.my

Abstract. Thermal comfort is a mean of being satisfied with the thermal environment. A study was conducted in the field to assess thermal comfort perception for the residents in a naturally ventilated residential building in Malaysia. The objective of this study was to find out whether the residents are thermally comfortable in buildings without the use of air-conditionings. The result showed that the residents are generally satisfied (i.e. comfortable) with the thermal environment, even though their thermal sensation was outside the central three categories of ASHRAE standard, which reflects their acclimatization to the local hot and humid climate.

1. Introduction

Thermal Comfort is defined as "the condition of mind that expresses satisfaction with the thermal environment and is assessed by subjective evaluation" [1]. As people spend most of their time in buildings, the indoor environments are needed to be more comfortable and healthy [2]. Psychological, physical, and social factors are involved when studying thermal comfort [3]. They can be categorized into three groups, namely; environmental factors (i.e. air temperature, relative humidity, air movement, and mean radiant temperature), individual factors (i.e. metabolic rate/activity and clothing), and contributing factors (i.e. Acclimatization, Experiences and Expectations, food and drink, body shape and subcutaneous fat, age and gender, and state of health). Due to variations in these factors, it is difficult to satisfy everybody in a space [1]. Therefore, a comfort zone is "the range of conditions where at least 80% of the people would feel comfortable" [4].

Natural ventilation is an attractive passive alternative that can provide and maintain a comfortable and healthy indoor environment. Besides, it helps to alleviate the associated problems with air-conditioned buildings and reducing energy consumption [5]. The objective of the current study is to investigate residents' perception of thermal comfort in a naturally ventilated residential building in Malaysia.

2. Literature review

Thermal comfort is more than just an air temperature. Psychological, physical, and social factors are involved when studying thermal comfort [3]. Current standards for thermal comfort are essentially based on either heat balance model or adaptive model [6]. The heat balance model is based on comprehensive analyses of the body thermal physiology by means of assuming a controlled steady-state conditions and a high accuracy for the six parameters of thermal comfort. However, the adaptive model depends on the active relationship between the occupant and the surrounding environments

Content from this work may be used under the terms of the Creative Commons Attribution 3.0 licence. Any further distribution of this work must maintain attribution to the author(s) and the title of the work, journal citation and DOI.
Published under licence by IOP Publishing Ltd

IOP Conf. Series: Materials Science and Engineering **401** (2018) 012009 doi:10.1088/1757-899X/401/1/012009

based on the concept that people usually react to restore their comfort levels whenever a discomfort condition existed.

Thermal comfort in hot climates was achieved with high temperatures. For example, in non-air-conditioned residential buildings in Kota Kinabalu city [7], a comfort temperature of nearly 30 °C was obtained with an acceptable range of 27 – 32.5 °C. Additionally, the neutral temperature for university students in their accommodations was 28.9 °C, while the comfort range was 27.6 – 30.1°C [8].

Studies in hot-humid climate for naturally ventilated buildings indicate that with higher indoor air velocity, an increase in the comfort temperatures voted by respondents is observed [1,6]. Additionally, moderate air velocity (>0.25 m/s) in the range 32-40°C reduce both thermal discomfort and skin moisture [9]. As a result, a study proposed lower limit for air velocity for three ranges of operative temperature; 0.4 m/s for 24-27 °C, 0.41-0.8 m/s for 27-29 °C and >0.81 m/s for 29-31 °C [10].

3. Methodology

A quantitative approach was selected for this research using questionnaire survey on thermal comfort. The thermal environment assessment was based on residents' votes using seven-point scales [1,11,12] including thermal sensation (TS) scale (ASHRAE scale), comfort perception (TC) scale (Bedford scale), votes of preference, direct votes of acceptability (Table 1).

Table 1. Rating scales of the survey.

ASHRAE scale		Bedford scale		Preference		Acceptability
+3	Hot	+3	Much too warm			
+2	Warm	+2	Too warm			
+1	Slightly Warm	+1	Comfortable warm	+1	Warmer	Acceptable
0	Neutral	0	Comfortable	0	No change	
-1	Slightly Cool	-1	Comfortable Cool	-1	Cooler	Not Acceptable
-2	Cool	-2	Too Cool			
-3	Cold	-3	Much too cool			

3.1. Case Study

N-Park, a high-rise residential building, was selected for this study due to: (1) Building's layout (i.e. single loaded corridor type) can provide the optimum opportunity for cross natural ventilation. (2) The orientation to north-south represents the recommended in Malaysia. (3) All the flats have the same orientation to the south, which provides the same conditions for all participants. It is located on the east coast of the Penang Island, Malaysia. It consists of four blocks (Figure 1) and has a total number of 988 flats. Each flat has a built-up area of 700 sq. ft., and comprises a master bedroom, two children bedrooms, living room, kitchen and two bathrooms. The design of all flats relies on natural ventilation. Even though, the owners installed air conditioners individually.

3.2. Population and sample size

Cochran's sample size equation for continuous data was used [13]. Based on this equation, for a population of 988, which represent one participant in each flat of all the 988 flats that involved in this study, the required sample size is 105. For 70% return rate [8], the required questionnaires to be distributed are 150 questionnaires.

3.3. Data Collection

Data collection lasted from 14th of February to 11th of March 2016. All the questionnaires were distributed from 9.00 am to 9.00 pm to the residents in their homes. Researcher effort was made in order to ensure getting participants with almost equal percentage of men and women with a range of ages and from different floor levels. Once the collected questionnaires were found not enough to achieve the required sample, another 50 questionnaires were distributed.

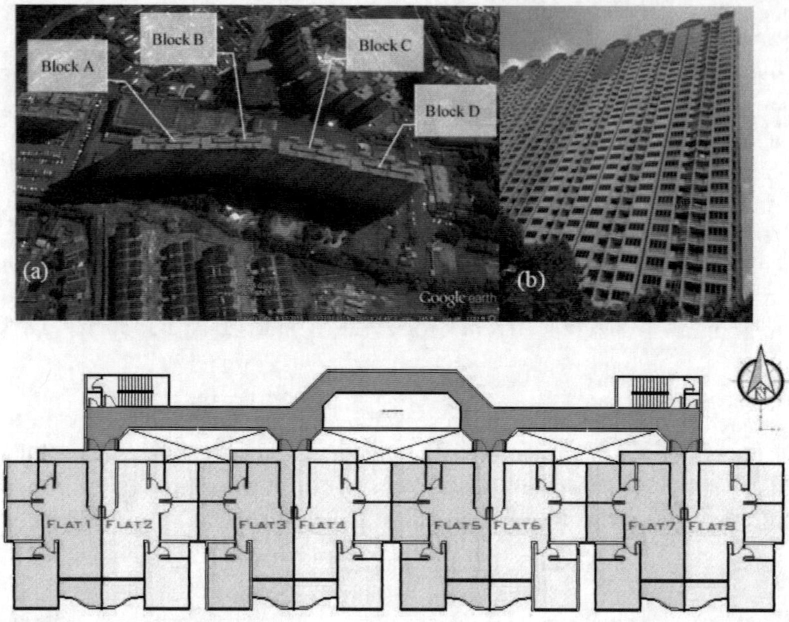

Figure 1. Case study (a) Building's blocks (source: Google earth), (b) Part of the south elevation (c) Block layout.

4. Results and discussion

From the 200 distributed questionnaires, 116 were collected and 106 were valid. The percentage of female to male is 38% to 62%.

Figure 2 shows respondents' votes for their thermal perception using ASHRAE scale (TS). It can be said that their votes clearly increased on the warm side of the scale. According to ASHRAE standard [1], the thermal condition may consider acceptable where at least 80% of the occupants voted within the central three categories (i.e. Slightly cool, Neutral, Slightly warm). The total respondents who voted within the three central categories is 77.3%. This indicates that the thermal condition considers not acceptable for the occupants. However, when respondents voted for their thermal comfort on Bedford scale (TC) (Figure 3), noticeably respondents' votes increased within the three central categories to 89.7% with an increase of 12.4% compared to ASHRAE scale.

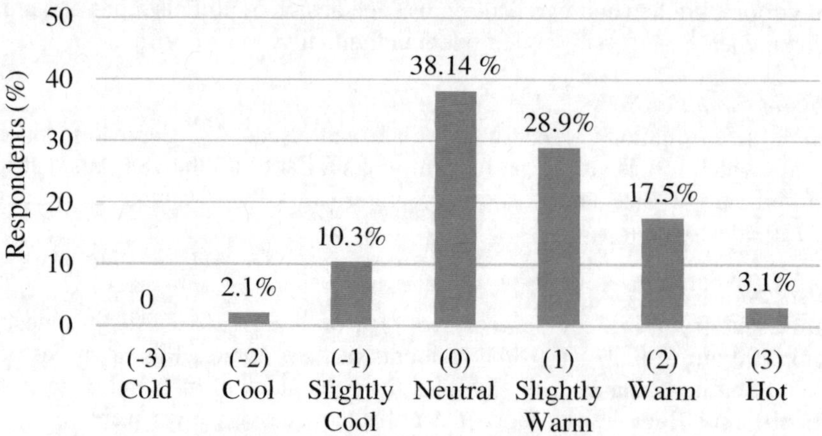

Figure 2. Thermal perception using ASHRAE scale (TS).

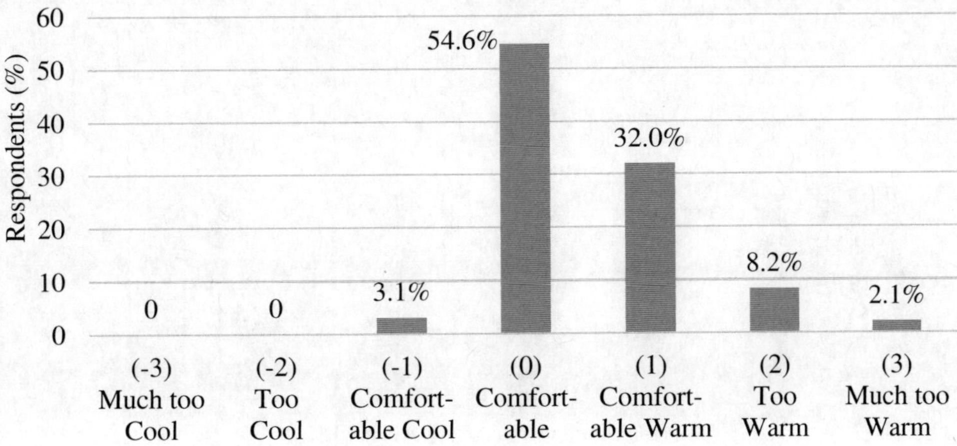

Figure 3. Comfort perception using Bedford scale (TC).

Similar results were found in studies conducted in hot and humid climates [8,14,15]. They mentioned that comfort assessment does not only depend on thermal sensation, but also on other non-physical factors (e.g. psychological factors such as expectation and emotional background). Evidence can be found in Table 2 (a cross frequency matrices of thermal sensation and comfort perception scales). From the 17.5% respondents who voted warm on thermal sensation scale, nearly two third (i.e. 11.3%) voted comfortable and comfortable warm.

Table 2. Cross frequency matrices of thermal sensation and comfort perception.

Thermal sensation (ASHRAE)		Comfort perception (Bedford scale)					
		Comfortable Cool	Comfortablc	Comfortable Warm	Too Warm	Much too Warm	Total
Cool	Total%	1.0%	0.0%	1.0%	0.0%	0.0%	2.1%
Slightly Cool	Total%	2.1%	7.2%	1.0%	0.0%	0.0%	10.3%
Neutral	Total%	0.0%	33.0%	4.1%	1.0%	0.0%	38.1%
Slightly Warm	Total%	0.0%	10.3%	17.5%	1.0%	0.0%	28.9%
Warm	Total%	0.0%	4.1%	7.2%	5.2%	1.0%	17.5%
Hot	Total%	0.0%	0.0%	1.0%	1.0%	1.0%	3.1%
Total	Total%	3.1%	54.6%	32.0%	8.2%	2.1%	100.0%

Data from the comfort survey was used to derive a regression of thermal sensation vote on reported comfort perception vote (Figure 4). The obtained linear regression equation ($r^2 = 0.3776$) is:

$$TC = 0.4522\ TS + 0.2497$$

In this equation, for (TS) vote of -0.55, the (TC) is zero, which indicates that the individual is at the (TC) of neutral state. This demonstrates that respondents normally will feel comfortable in a cooler environment. For TC votes of -1 and +1, the corresponded TS votes are -2.76 and +1.66 respectively. The higher TS votes on the cooler side are being rated as more comfortable than the warm side. This also proves that respondents can be thermally comfortable with TS votes outside the central three categories that suggested by (ASHRAE) for comfort conditions.

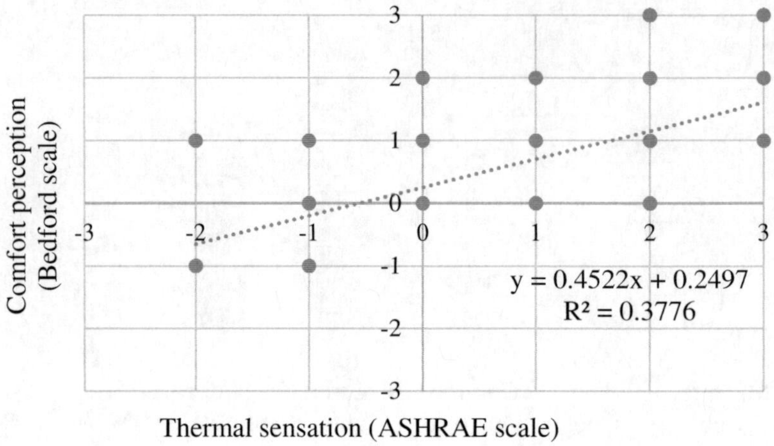

Figure 4. Liner Regression of thermal sensation and comfort perception scales.

Additionally, when residents were asked about their preference to change the thermal environment, 63.2% voted to be cooler, while just 34.9% voted for no change, Figure 5 (a). People may find an environment thermally comfortable even if they still preferred to be cooler, which supports the earlier findings with linear regression. This is also in line with Toe & Kubota finding that people in hot climates can adapt to higher neutral temperatures, yet they still prefer cooler conditions [6]. Furthermore, Figure 5 (b) shows that as much as 84.9% of the respondents accepted the environment. This further confirms the findings from comfort perception Bedford scale. It also proves the concept of acclimatization and how the human body in hot climates may develop a tolerance to high temperature and feel comfortable compared to people in other climates.

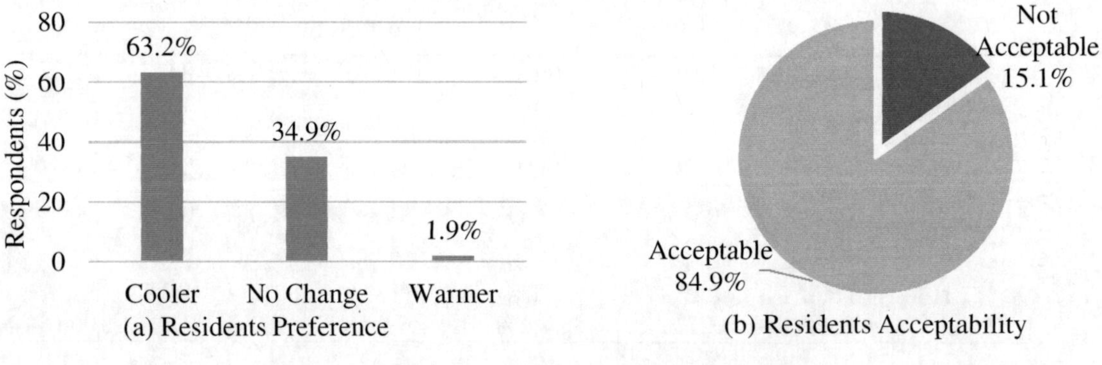

Figure 5. Residents Preference and Acceptability.

5. Conclusion

This study was conducted to assess thermal comfort perception for the residents in naturally ventilated residential building in Malaysia. The main results showed that residents are generally satisfied (i.e. comfortable) with the thermal environment in the naturally ventilated building, even though their thermal sensation was outside the central three categories of ASHRAE standard, which reflect their acclimatization to the local hot and humid climate.

Acknowledgement

The authors would like to thank Sana'a University, Universiti Sains Malaysia and FRGS (203/PPBGN/6711516) grant for the financial support for this research.

References

[1] ASHRAE 2004 *ANSI/ASHRAE STANDARD 55-2004 Thermal Environmental Conditions for Human Occupancy* (Atlanta, GA: American Society of Heating, Refrigerating and Air-Conditioning Engineers, Inc.)

[2] Daghigh R 2015 Assessing the thermal comfort and ventilation in Malaysia and the surrounding regions *Renew. Sustain. Energy Rev.* **48** 681–91

[3] Nicol 1993 *Thermal Comfort - A Handbook for Field Studies Toward an Adaptive Model* (UK: University of East London)

[4] Koenigsberger O H, Ingersoll T G, Mayhew A and Szokolay S V 1974 Manual of tropical housing and building. Climatic design 1–335

[5] Montazeri H, Montazeri F, Azizian R and Mostafavi S 2010 Two-sided wind catcher performance evaluation using experimental, numerical and analytical modeling *Renew. Energy* **35** 1424–35

[6] Toe D H C and Kubota T 2013 Development of an adaptive thermal comfort equation for naturally ventilated buildings in hot–humid climates using ASHRAE RP-884 database *Front. Archit. Res.* **2** 278–91

[7] Djamila H, Chu C-M and Kumaresan S 2013 Field study of thermal comfort in residential buildings in the equatorial hot-humid climate of Malaysia *Build. Environ.* **62** 133–42

[8] Wafi S R S 2012 *Thermal Comfort of Students' Accommodations in Universiti Sains Malaysia* (Ph.D. thesis. Universiti Sains Malaysia)

[9] Nicol J F 2009 An analysis of some observations of thermal comfort in Roorkee, India and Baghdad, Iraq *Ann. Hum. Biol.*

[10] Cândido C, Lamberts R, de Dear R, Bittencourt L and de Vecchi R 2011 Towards a Brazilian standard for naturally ventilated buildings: guidelines for thermal and air movement acceptability *Build. Res. Inf.* **39** 145–53

[11] En 15251 2007 *Indoor environmental input parameters for design and assessment of energy performance of buildings addressing indoor air quality, thermal environment, lighting and acoustics* vol 3 (Brussels: BSI)

[12] McCartney K J and Nicol F J 2002 Developing an adaptive control algorithm for Europe *Energy Build.* **34** 623–35

[13] Bartlett J E, Kotrlik J W K J W and Higgins C 2001 Organizational research: Determining appropriate sample size in survey research *Inf. Technol. Learn. Perform. J.* **19** 43–50

[14] Wong N H, Feriadi H, Lim P Y, Tham K W, Sekhar C and Cheong K W 2002 Thermal comfort evaluation of naturally ventilated public housing in Singapore *Build. Environ.* **37** 1267–77

[15] Feriadi H and Wong N H 2004 Thermal comfort for naturally ventilated houses in Indonesia *Energy Build.* **36** 614–26

IOP Conf. Series: Materials Science and Engineering **401** (2018) 012010 doi:10.1088/1757-899X/401/1/012010

ARCHITECTURAL VALUE FOR URBAN TOURISM PLACEMAKING TO REJUVENATE THE CITYSCAPE IN JOHOR BAHRU

[1]Noraslinda Abdul Rahman, [2]Norhazliza Halim, [3]Khalilah Zakariya

[1]Center for the Study of Built Environment in the Malay World (KALAM), Faculty of Built Environment, Universiti Teknologi Malaysia, Skudai 81300, Malaysia. (noraslinda.ar@utm.my)
[2]Centre for Innovative Planning and Development (CiPD), Faculty of Built Environment, Universiti Teknologi Malaysia, Skudai 81300, Malaysia. (norhaz@utm.my)
[3]Kulliyyah of Architecture & Environmental Design, International Islamic University Malaysia, Kuala Lumpur 53100, Malaysia. (khalilah@iium.edu.my)

Abstract. This study aims to investigate the social significance of urban placemaking in shaping the architectural values of the cityscapes for sustainable tourism. The Johor Bahru City Centre was developed with explicit intention to have public open spaces that can encourage tourism activities in the urban area. Cultural and natural resources are essential to make the tourism industry grow successfully. Historical and archaeological sites also become a part of tourist attraction such as museums, state landmarks, parks, sculptures, monuments and others. In this respect, the architectural values of the city are the tangible assets that the city has, while the cultural and natural values are the intangible assets that can be captured from places that shape tourist perception. Placemaking supports the concept of generating places that improve the relationship between users and space, by increasing the sense of place. This is where the values of architecture must be preserved in order to maintain the identity of the region. Based on an extensive review of the literature and empirical works, this research examines Johor Bahru by extracting its similarities and differences with cities in Singapore and Australia, to derive at a better understanding of the significance of the architectural value and urban tourism placemaking in rejuvenating the city values attached to physical spaces. The benchmarking study analyses the values of architecture and discusses the advantages resulting from various architectural characters in tourism. These relationships are essential to understanding the role of architectural value for urban tourism placemaking in rejuvenating the city. This study found that there is a complex nature of the generations of architectural elements through placemaking.

Keywords: *Architecture in tourism; placemaking in tourism; sustainable city tourism, urban tourism placemaking, cityscape rejuvenation.*

Content from this work may be used under the terms of the Creative Commons Attribution 3.0 licence. Any further distribution of this work must maintain attribution to the author(s) and the title of the work, journal citation and DOI.
Published under licence by IOP Publishing Ltd

1. Introduction

Architecture is a physical form that plays a significant role in urban tourism placemaking. Architectural elements such as buildings, monuments, sculpture, park, landscapes and public spaces contribute to the identities of places. The presence of these elements as part of the urban fabric makes cities as one of the most desired attractions of tourism industry. According to Edwards, Griffin and Hayllar (2008) in Scerri, Edwards and Foley (2016), physical form as defined by architecture refers to space with aesthetic value that can help people to recognize and feel a sense of place. It will also help to "pull" tourists to the area through the various focal points that attract visitors' attention and offer them a rich experience (Krolikowski and Brown, 2008). In this sense, the values of architecture can be seen to have a direct influence on tourism motivation. Tourism motivation is essentially the "push" and "pull" factors associated with people's desire to go on a vacation (push factor) and choice of destination (pull factor). Throughout the world, architecture has always had a strong influence in tourism. For instance, as observed in mass tourism, historical buildings, palaces and monuments such as those found in Europe and Asia, have continuously received a high number of tourist arrivals from all over the world. Great architecture can be an instrument to enhance a destination image, or even put a less known destination on the map. According to Scerri, Edwards and Foley (2016), buildings can make places easy to identify. The imagery of the architectural becomes synonymous to the image of the destination. The authors further added that historic buildings can capture celebrations, events and promote renowned people, while museum is a building that holds and displays cultural products and historical artifacts that serves as an educational reference point for the community and tourists. The role of architecture in urban tourism encompasses a wide field, from the architectural infrastructure needed to transport and host tourists, to the physical attractions that tourists visit. Tourism sells a beautiful experience. The tourist experience takes place in a setting that will always be remembered. Architecture is a major component of the city's setting, and therefore its values are essential in the making of successful urban tourism.

Tourism placemaking is a process that generates destination image and it is very crucial to enhance repeat visitations by visitors (Razali and Ismail, 2014). The importance of the placemaking concept is to make a place have a distinctive identity, life, sustainability, resilience, reflect on life and community involvement in its planning and development. Urban tourism placemaking is a tool to transform the physical forms of the city into places that tourists desire to spend their time and money. In this process, spaces are converted into meaningful places for the local community and the tourists, which eventually becomes part of the tourist motivation. Placemaking supports the concept of generating places that improve the relationship between users and space, by increasing the sense of place. In the placemaking process, emphasis is given to both the local community's needs and the tourists' needs. The balanced approach is integral part of a city's development in ensuring that the place can be sustained for a long time because essentially the caretakers of the city are the locals themselves. This is where architecture plays an important role in maintaining the identity of the cityscape that needs to be preserved. The cityscape is a view or picture of the place that simultaneously describes the characteristic and appearance of a city. The city represents its people, as much as the people represent their city. The cityscape reflects the social space and the overall destination image, and this can influence the time spent by tourists to do activities in the area. Embarking from this notion, this study aims to investigate social significance of urban placemaking in shaping sustainable tourism especially in architectural values of the cityscapes.

2. Architecturalvalue for Urban Tourism Placemaking

Urban Tourism Placemaking Concept

The concept of tourism placemaking is a planning and development approach in constituting an image and sense of destination to enrich the traveler's experience. Tourism placemaking is a value-added process that enhances destination identities to visitors and tourists through assets of the local communities, such as locality, history, culture and environmental values (Canter 1977; Gunn 1988; Potteiger and Purinton 1998; Bell 1999; Zakariya, Mohyuddin and Yaman, 2007). The concept of tourism placemaking is a destination development that reinforces on the characters, identities and meanings of the attractions displayed and offered by the community as tourism products that offer travelers an interesting experience (Smith 1995; Pratiwi, Soedarmadji and Yanindaputri 2010; Notten 2011). Therefore, in the context of this study, the concept of tourism placemaking is a planning and development approach in establishing identity and sense of place in a destination, to give travelers a unique experience and surroundings (Razali et al., 2017). The presence of physical elements in a place, such as built forms and natural forms, shape the distinctiveness of the place. All places stand on a specific geographical location. Therefore, how the places are built are usually in response to their local climate, local culture and local resources. Lynch (1960) suggests that what sets a place a part from one another defines its identity. This identity can sometimes be visually seen from its built forms such as the buildings, monuments and city spaces. Nonetheless, part of this identity is culturally defined from how the place is inhabited, used, adapted and becomes assimilated in the people's way of life. This study makes references to the town centre of Singapore and the city of Adelaide in Australia as a starting point to study the similarities and differences with Johor Bahru to deliver better understanding about rejuvenating the city through architecture and urban tourism placemaking attached to physical spaces. Physical space is important in determining the identity of the city, as it is the point of social and cultural discovery and community to interact.

Placemaking Theory

In theory, placemaking is the transformation of a space into place (Razali et al., 2017). The original concept of placemaking is the planning and management of public space through the community (Wykoff, 2014). In the tourism context, space and place create tourist attractions because tourist attraction is important to generate tourist presence, business, economic activity and mobility where the host and organization need to think about how to attract people to come (Hultman 2007; Jensen 2007; Ooi, Hakanson and Lacava et al. 2014; Ertan and Yakup, 2015). In terms of strategic destinations, it is a geographical brand of attraction, culture and environment for marketing strategy (Gunn 1988; Cooper and Michael Hall 2008; Hultman and Hall 2012). The settings of the destination combined with the travel behavior of the tourists shape the tourist experience (Ryan, 1995). Figure 1 shows the placemaking concept between space theory and place theory, thus creating tourism destination.

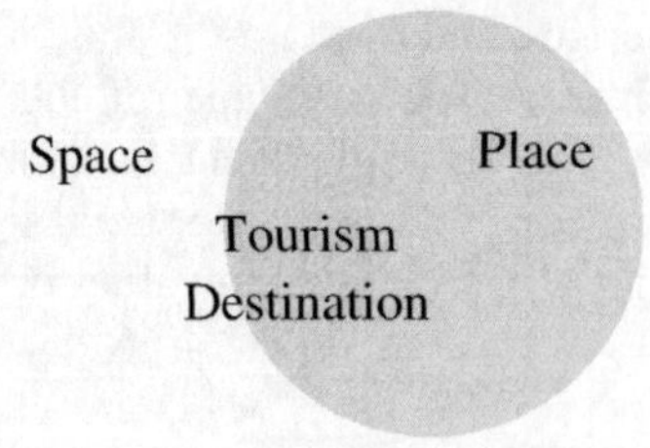

Figure 1: Placemaking Concept in Tourism
Source: (Relph 1976; Tuan 1977; Gunn 2002)

Cityscape Rejuvenation That Involve Architectural Value

Transformation of space into place is very important to rejuvenate a place. Humans need a space for them to continue their life as stated by Greenbie (1992) that as a civilized society, people need a shared space uninterrupted and invaded by one another. Place is the result of space that is inhabited or interacted by people, and the place usually has a fixed location and certain characteristics that give its own character and value (Relph, 1976). With reference to the city of Adelaide, located in South Australia, we can see an example of how a space is transformed in to place where the street architecture is combined with placemaking concept to rebrand the town. Adelaide, which is the capital city, started the placemaking activities a few years ago by transforming their parking spaces into a place where the community can meet and do it some activities, like food vendors, musicians and street shops. Adelaide uses the placemaking approach as a tool to engage the public in new ways. For example, they maintain the city form and heritage buildings while improving the quality of street environment (Adelaide City Council, 2014). According to the Adelaide's local area development plans under Development Act 1993, they promote the use of placemaking and shared spaces around the city with several strategies like a building façade fronting street frontages, access way, driveway or public spaces should be composed with an appropriate scale, rhythm and proportion which respond to the use of the building, local character and surrounding building. Besides, building, advertisements, site landscape, street planting and paving should be integrated with coordinated appearance to enhance the cityscape urban environment (Adelaide City Council, 2014). Thus, it shows clearly that architectural value is very important for the rejuvenation of a city to increase the arrival of tourists.

In this respect, architectural value represents the tangible benefit in urban tourism placemaking, while cultural and natural values are the intangible benefit that can be captured from places that shape tourist perception. With reference to the Singapore Tourism Board (STB) website, STB together with various government agencies, precinct associations and private stakeholders, convene in a coordinated effort to spearhead, develop and implement various placemaking initiatives, such as festivals, marketing initiatives and infrastructure improvements, with the aim to improve visitor experience and inject vibrancy to bring the precincts to life. Singapore enhances its architectural value and use the placemaking approach to shape the image and sense of place that influence the tourist experiences. For instance, they create Cultural Precincts like Chinatown and Little India by improving the street market, heritage architecture building and food street. They also propose many festivals to provide a worthwhile experience for travelers. Additionally, they also create Lifestyle Precincts like Orchard Road and Sentosa Harbourfront. Each precinct has unique characteristic through differentiated leisure events that attract visitors and increase spending (Singapore Tourism Board, 2018). These reference studies prove that Singapore and Adelaide emphasize their architectural, cultural and natural values to rejuvenate their city in order to enhance tourist motivation.

3. Architecture Value in Johor Bahru Cityscape

Johor Bahru is capital city of Johor state, located in the southernmost part of the Malaysia Peninsula, adjacent to Singapore with a population of half million people making it second largest city in Malaysia (Statistic Department, Malaysia, 2010). Johor Bahru has its own unique capability to attract domestic and international visitors, according to Tourism Malaysia Statistic 2017 that indicates 7.01 million tourists stayed overnight in Johor with the largest percentage is in Johor Bahru as much as 69%. The tourism length of stay is a variable of key factor for any tourism destination due to its crucial effect on overall tourism expenditure (Economics, Thrane and Science, 2015).

IOP Conf. Series: Materials Science and Engineering **401** (2018) 012010 doi:10.1088/1757-899X/401/1/012010

Figure 2 shows the Johor Bahru cityscape that has more elements of architecture, which makes this place so unique and rich in history.

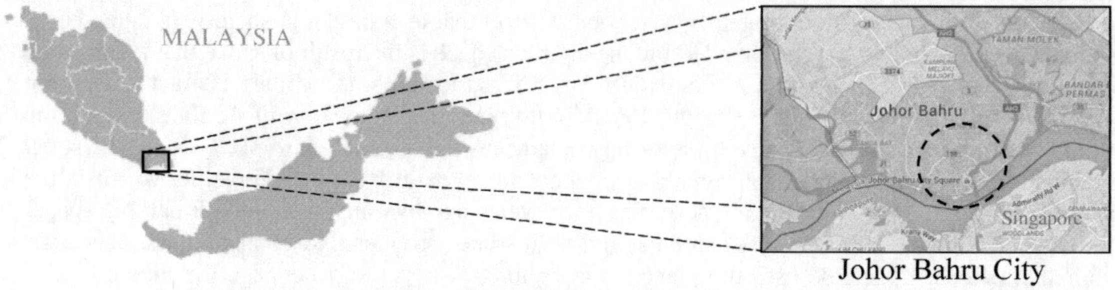

Figure 2: Johor Bahru Location.

Figure 3: Johor Bahru Building Cityscape

Source: (Google Map andhttps://graphicriver.net/item/johor-bahru-malaysia-skyline-with-gray-buildings-blue-sky-and-reflections/20531792)

Johor Bahru cityscape has challenges due to some public spaces that do not present a sense of invitation to attract people to sit and rest due to congestion even on the pedestrian walkways, lack of proper sit-outs, pollution and absence of well-landscape open space and street light (Rahely, Najafpour and Lamit, 2013). It can be seen that most of the weakness are caused by lack of architectural values such as street architecture, landscape architecture and others. Johor Bahru's interesting places and importance of architectural value for urban tourism placemaking have potentials to rejuvenate the cityscape in the town. Johor Bahru's rejuvenation involves urban regeneration process of remaking the place. In this process, regeneration initiatives are planned to enhance the physical conditions of places, increase economic growth and environmental sustainability in order to facilitate higher quality of people life. Rebranding the city is able to attract local and foreign tourist to come to a place as well as increase tourism activity in Johor Bahru. However, it is difficult to discuss tourism without discussing architecture because architecture is a commodity of tourist consumption and objectified cultural capital (Scerri, Edwards and Foley, 2016). The Johor architecture is formed from a combination of Dutch, Malay, Chinese, Indian, Portuguese, and British during the various eras. Ethnic diversity produces uniqueness for this city. The place rich in cultural and natural resources has the potential to make tourism to grow successfully. Other than that, the historic and archaeological sites also become part of tourist attraction, such as museum, landmark state, parks, sculpture, monuments and others.

4. Johor Bahru Cityscape and Its Architecture

The value of architecture plays an important role in rejuvenating the Johor Bahru cityscape. Based on site observation, Johor Bahru cityscape consists of diversity architecture such as royal architecture, religious architecture, state administrative building architecture, cultural architecture, modern architecture, street architecture, landscape architecture and others. Beautiful and wonderful architecture can be seen in the royal building which symbolizes the superiority status of Johor. For instance, Sultan Abu Bakar Mosque is a Johor's state mosque as beautiful example of Victorian fusion architecture and perched atop a hill with sweeping views of the Johor Strait and across to Singapore. Furthermore, royal architecture can be observed closely at Abu Bakar Royal Palace surrounded by several hectares of beautifully manicured grounds was completed in 1866. In 1990, it was reimagined as a public museum with an impressive collection of royal artefacts. The value of Johor architecture is also absorbed in the official building of the state administration like Sultan Ibrahim Building as a monument of western and eastern architecture, a symbol where traditional and modern civilization blended harmoniously together. In addition, the building was the earliest skyscrapers and stood the tallest amongst the buildings in Johor Bahru until the 1970s and as symbol of the King's visionary plans for Johor (Tourism Johor Website, 2018). Another historic building is Johor Bahru High Court built in 1900 shows the features of Italian Renaissance Architecture because Johor was brought under British control when the court was built. Other than that, Johor Bahru Police Station influenced by architecture from England with blue and white furnishing (Tourism Johor Website, 2018). Near the police station there is a historic building that connecting two neighboring countries Malaysia-Singapore, Johor Bahru Old Railway Station was launched in 1932 and has been intercity railway station that boasts western architecture in a very unique way. Now, the building opening as train museum and provides opportunities for travelers to gain a great architectural experience.

Johor Bahru cityscape formed by strong relationship among multiracial communities. It can be proven by existence of various religious architecture such as Sultan Abu Bakar Mosque for Muslims, Johor Bahru Old Chinese Temple for Chinese community was built in 19th century with combination of different Chinese group as Hokkien, Cantonese, Hainan, Teochew and Hakka which combine all the Chinese identity. Sri Raja Mariamman Temple Johor Bahru Immerse in rich culture of Hindu with over 100 years of history, the impressive Hindu temple was a huge contrast to nearby hotels and shopping malls. Next is the Johor Bahru Gurdwara Sahib Sikh Temple a most unique and exotic heritage of Sikh with witnessed in ancient time, most of the Sikhs were appointed to work as policemen and the bodyguards of royalty. Besides that there is also have India Muslim Mosque Located right at the junction of Jalan Duke and Jalan Dhoby featuring a silver and blue dome with two minarets aside, the mosque comes with white walls with blue-tinted glass, offering a modern ambience and convenient. There are also old buildings that are given a new life by maintaining the original architecture, Tiong Hua Chinese Heritage Museum is a newly-renovated museum that boasts four stories and stands prominently in the middle of the row of shop house. Before converting into a museum, it was previously owned by JB Tiong Hua Association as their base.

Johor Bahru cityscape aim to be a thriving and dynamic tourism destination by 2020 with Family Fun theme by providing world class recreational, tourism circuit and as southern tourism destination city of Malaysia (Iskandar Regional Development Authority, 2013). Hence, the cityscape is undergoing changes to achieve the vision and can be seen with the construction more tourist's area are built by targeting all segmentation including children, teens, adults and families. For example, JB crown also known as Laman Mahkota become as tourist attraction. The crown arch that complements the palace was actually built as a gift by the Sultan of Johor for the people (The Official Tourism Johor Website, 2018). The cityscape is enlivened with night atmosphere in the middle of the capital city, Johor Bahru Bazar Karat is a good place to buy cheap and rare items such as clothes, accessories, food, pets, antics and music products. The flea market has a history and many tourists come over to get close to local culture. The most amazing this street architecture is Johor Bahru Bazar Karat is surrounded by old and historic building, it provides travelers with the opportunity to appreciate the architecture at night. Other than that, the important trade history for this city is Segget River located in the heart of city and has a history or urban city development. Now, the segment area is undergoing revolution to become a fascinating place to attract more target segmentation. All the above examples clearly show that the architecture has its own value in forming urban tourism placemaking to rejuvenate the cityscape in Johor Bahru.This is where the importance of architecture must be preserve in order to maintain the identity of the region.

5. Discussion and Conclusion

This paper has discussed the role of architectural value for urban tourism placemaking in adding to the rejuvenation of place. This study demonstrates the complex nature of the generation of architectural elements through placemaking and the importance of sustaining the rich architectural characters of the city. Iconic architecture can make cities easily recognisable, where historic buildings capture the commemoration of famous events and people, and galleries and museums that hold and display cultural values and serve as places of focus and inspiration that bring communities together. In this paper, several placemaking strategies have been identified to rejuvenate the cityscape in Johor Bahru. First, the design of building can support places, for example, “green” building at the ground level, it is dynamically connected to the surrounding neighborhood, fostering street life and creating a strong sense of place. The area around the building can be enhanced by shaded structures and other amenities, making this a comfortable place and an integral part of the community. Architecture is important in urban tourism placemaking to rejuvenate the cityscape especially in Johor Bahru. Secondly, the improvement of streets as places can start with designing the street as comfortable and safe place for everyone. Squares and parks can become multi-use spaces for events, as has been implemented in Johor Bahru city by turning the parking area to popular night market. Third, there is a need to preserve and protect the city’s resources. Protecting the historic landmarks and places, unique qualities, and special traditions are crucial to attract visitors because tourism has an impact on the resources and puts stress and strain on infrastructure such as roads, landscape, water supplies, and public services. Fourth, the city needs to be able to position its uniqueness and branding by highlighting its sense of place and stories, and make the tourism site come alive by providing accurate, authentic information in an engaging and memorable way that makes the site come alive for the visitor. Finally, the tourism agencies and planners need to collaborate with partners and form alliances as an excellent way to leverage scarce financial and human resources is to form partnerships among historic sites, cultural attractions, organizations, governments, and businesses in the city. These collaborations can create multifaceted packages of traditional, cultural, and heritage tourism opportunities, thereby offering an enhanced experience to the traveller.

REFERENCES

[1] Bell, S. (1999). Landscape; Pattern, Perception and Process Second. London: E&FN Spon.

[2] Canter, D.V. (1977). The psychology of place, London: Architectural Press.

[3] Cooper, C. and Michael, H.C. (2008). Contemporary tourism: an international approach, Oxford: Goodfellow Publishers Ltd.

[4] Economics, T., Thrane, C., and Sciences, A. (2015). Tourist's at length of stay : The case of international summer visitors to Tourists' length of stay : the case of international summer visitors to Norway, (February). https://doi.org/10.5367/te.2012.0158

[5] Greenbie, B. B. (1982). Spaces: Dimensions of the Human Landscape, New York: Yale Press.

[6] Gunn, C. A. (2002). Tourism Planning: Basics, Concepts and Cases, Washington: Taylor & Francis.

[7] Hultman, J. (2007). Through the Protocol: Culture, Magic and GIS in the Creation of Regional Attractiveness. Tourism Geographies, 9(3), pp.318–336.

[8] Hultman, J. and Hall, C. M. (2012). Tourism place-making. Governance of Locality in Sweden. Annals of Tourism Research, 39(2), pp.547–570.

[9] Jensen O B (2007). Culture Stories: Understanding Cultural urban Branding. Planning Theory, 6(3), pp.211–236.

[10] Lynch, Kevin. The image of the city.Vol. 11. MIT press, 1960.

[11] Notten E (2011). Tourism and Place Making In Clanwilliam, South Africa. Wageningen University.

[12] Ooi C-S, Håkanson L and Lacava L (2014). Poetics and Politics of the European Capital of Culture Project. Procedia -Social and Behavioral Sciences, 148, pp.420–427.

[13] Placemaking. (2018, March 8). Retrieved from https://www.stb.gov.sg/about-stb/what-we-do/Pages/Placemaking.aspx

[14] Potteiger M and Purinton J (1998). Landscape Narratives First., Canada: JohnWiley & Sons, Inc.

[15] Pratiwi W D, Soedarmadji N and Yanindaputri P (2010). Place-Making for Tourism in RuralSettlements and the Institutional Collaboration. Arte-Polis 3 International Conference - Creative Collaboration and the Making of Place Place-Making, pp.373–382.

[16] Rahely, E., Najafpour, H., and Lamit, H. (2013). Public Places and Spaces and Social Urban Interaction (A Case Study of Johor Bahru, Malaysia). International Journal of Current Engineering and Technology, (January).

[17] Razali, M. K., Ahmad, H., Jusoh, H., and Choy, E. A. (2017). Place-making Dalam Agenda Pembangunan Pelancongan Place-making in Tourism Development Agenda. Place-Making Dalam Agenda Pembangunan Pelancongan, 1(1), 38–50.

[18] Razali, M. K., and Ismail, H. N. (2014). Tourism Place-Making at Tourism Destination from a Concept of Governance. IGCESH2014 Universiti Teknologi Malaysia, Johor Bahru, Malaysia 19-21 August 2014, (August), 172–177.

[19] Relph, E. (1976). Place and Placelessness, London: Pion Limited.

[20] Ryan, C. (1995). Researching tourist satisfaction.London: Routledge.

[21] Scerri,M. Edwards, D. andFoley, C. (2016). The value of architecture to tourism. The Value of Architecture to Tourism, (February).
[22] Smith, S.L.J. (1995). Tourism Analysis a Handbook First Edit., Harlow: Longman Group Limited.
[23] Tuan, Y.F. (1977). Space and Place: The Perspective of Experience First Edit., Minneapolis: University of Minnesota Press.
[24] Tourism Johor, (2018). Explore Johor. Retrieved from http://tourism.johor.my
[25] Tuğçe, E& Yakup, E (2015). An Integrated Approach to Urban Preservation A Suggestion To Combine Two Methods of Urban Regeneration In Order To Achieve a Further Develop Urban Historic Center Preservation. GSTF Journal of Engineering Technology (JET), 3(3), pp.1–7.
[26] Wyckoff, M.A. (2014). Definition of Placemaking: Four Different Types. Planning and Zoning News. Retrieved fromhttp://pznews.net/media/
[27] Zakariya K, Mohyuddin A & Yaman M (2007). Refining Tourist's Place Experience through Placemaking: Concepts and Correlations. The International Journal of Diversity in Organisations, Communities and Nations, 7(4), pp.249–257.

IOP Conf. Series: Materials Science and Engineering **401** (2018) 012011 doi:10.1088/1757-899X/401/1/012011

Effects of roof pitch gradient and material to harvested rainwater quality

Lai, Y.H.[1], Ahmad, Y.[1], Yusoff, I.[2], Bong, C.W.[3,4], Kong, S. Y.[1]

[1] Department of Architecture, Faculty of Built Environment, University of Malaya, 50603 Kuala Lumpur, Malaysia.

[2] Geology Department, Faculty of Science, Universiti Malaya, 50603 Kuala Lumpur, Malaysia.

[3] Institute of Biological Sciences, University of Malaya, 50603 Kuala Lumpur, Malaysia.

[4] Institute of Ocean and Earth Sciences (IOES), University of Malaya, 50603 Kuala Lumpur, Malaysia

k_sengyeap@um.edu.my

Abstract. Rainwater harvesting has untapped potential in domestic usage due to the lack of awareness and design guidelines to building sustainable roofs. Studies suggest that combinations of different roof pitch gradient and materials will affect the runoff quality. The level of contamination and contents found in the harvested rainwater will require different ways for water treatments prior to usage. The objective of this research is to identify design criteria of a building roof affecting the roof runoff quality in Kuala Lumpur. Quantitative analysis is done based on the rainwater collected from case studies with combinations of different roof pitch and materials to identify the contents and level of contamination in each of the collected sample. The result observed the best runoff quality from clay tile roof of 45°, metal roof of 15° and polycarbonate roof of 15°; whilst reinforced concrete (RC) flat roofs harvested the most contaminated runoffs. The study shows that all roof runoff quality is not up to potable standard though can be used for indoor non-potable use with minimal treatment.

1. Introduction

Water scarcity has become the top global crisis causing pressure on the resources that are vital for human survival. [1][2] Rainwater harvesting (RWH) has emerged as one of the best solutions to relieve urban issues such as flash flood and high demand for treated water. [3] Other than ground-surface storm water collection, rainwater is more commonly collected from above-ground rooftop catchment areas which are potentially less polluted and easily treated for domestic use. [4][5][6]

Various studies have identified the relation between roof material and runoff quality with different outcomes due to the difference in climate, context and variation of each material in different localities [7]. Ferreny et al. (2011) suggested that flat gravel roofs harvested the most contaminated rainwater. [8] The pattern observed from this research is that sloping smooth roof will result in the best quality and quantity of roof run-offs. Studies conducted by Lee et al. (2012), Olaoye et al. (2012) and Despin et al. (2009) showed similar results with best harvested runoff quality from steel and aluminium roofings; [3][7][9] while Simmons et al. (2001) discovered that lead contamination was found higher in RWH

Content from this work may be used under the terms of the Creative Commons Attribution 3.0 licence. Any further distribution of this work must maintain attribution to the author(s) and the title of the work, journal citation and DOI.
Published under licence by IOP Publishing Ltd

IOP Conf. Series: Materials Science and Engineering **401** (2018) 012011 doi:10.1088/1757-899X/401/1/012011

systems with lead or galvanized iron as part of the system. [10] Chang et al. (2004) suggested that the reduction of zinc used in constructing RWH system and components is essential to reduce zinc pollution. [11] Abbott et al. (2007) conducted a research on the micro-biological quality of roof –harvested rainwater in relation to the awareness of the respective residents on rainwater and roof maintenance. [12] As important as the awareness of end-users to maintain their roofs and to treat the harvested rainwater, the cost and effort of doing so can be reduced if the roof design and material are optimized in the beginning.

2. Methodology

The research was done within University Malaya campus in Kuala Lumpur to ensure consistency of the microclimate and urban activities which will affect the rainwater quality. Nine (9) roofs were selected as case studies (three clay tile roofs, two metal roofs, two polycarbonate roofs and two RC flat roofs) based on deciding factors such as the roof material and pitch gradient as well as controlling factors such as building height at one storey and availability of drip-off collection point.

Rainwater drip-off from the roof eaves of the nine selected case studies and one ambient rain sample (controlled sample) were collected during a drizzling rain event (wet season). Approximately 400 ml of rainwater for each case study was collected in a laboratory container for physical tests and IC test; while another 10 ml was acidified on site with 3 drops of diluted hydrochloric acid for ICP test. Due to the low visibility and time restriction during on-site data collection, physical tests were conducted in the laboratory as soon as data collection is completed.

Table 1. Tabulation of sample coding based on roof attributes.

Roof Type	**Clay Tile**			**Metal**		**Polycarbonate**		**RC Flat Roof**		**Ambient Rain**
Pitch	15°	30°	45°	15°	30°	15°	30°	<5°	<5°	-
Code	C1	C2	C3	M1	M2	P1	P2	RC1	RC2	A1

Table 2. Tabulation of precipitation data for sample collection.

Parameter	**Units**	**Method / Test**
Physical		
pH	upH	pH benchtop meter
Conductivity	µs/cm	Portable multiparameter meter
Total suspended solid (TSS)	ppm	
Salinity	psu	
Chemical		
Ions (F-, Cl-, NO_2-, NO_3-, $SO_{4\text{-}2}$)	mg/L	Ion chromatography test (IC test)
Heavy Metal (Al, Zn, Fe)	mg/L	Inductively coupled plasma test (ICP test)
Hardness (Ca, Mg)	mg/L	

The results of each parameter were plotted into graphs, where performances were compared between samples and National Water Quality Standards (NWQS) for Malaysia. The discussion of each graph will lead to a spider chart marked for each sample for a clearer picture of its general attributes in terms of rainwater harvesting efficiency and performance.

3. Results and Discussion

Table 3. Tabulation of results.

Parameters	Units	C1	C2	C3	M1	M2	P1	P2	RC1	RC2	A1
Physical											
pH	upH	6.91	7.18	7.15	6.20	5.51	5.57	5.21	7.35	7.51	7.18
Conductivity	µs/cm	80.76	57.73	56.46	11.81	17.82	20.11	37.42	105.6	117.0	10.44
Total Dissolved Solid	mg/L	39.64	29.03	27.64	6.30	9.31	10.55	18.68	52.55	58.22	5.57
Salinity	psu	0.09	0.08	0.08	0.06	0.06	0.07	0.07	0.103	0.1	0.06
Main Ions											
Nitrate, NO_3-	mg/L	5.00	4.53	2.89	1.68	3.23	2.29	5.81	1.04	0.31	0.90
Sulphate, SO_4-	mg/L	4.55	3.07	1.98	1.89	2.17	1.70	3.31	3.86	3.44	0.81
Hardness											
Calcium, Ca	mg/L	13.71	9.19	9.62	0.31	0.42	0.41	3.56	18.71	22.49	0.31
Magnesium, Mg	mg/L	0.12	0.25	0.17	0.01	0.04	0.01	0.05	0.31	0.17	0.01
Heavy Metal											
Aluminium, Al	mg/L	0.13	0.14	0.15	0.47	0.49	0.45	0.43	0.11	0.17	0.47
Zinc, Zn	mg/L	0.08	0.04	0.02	0.04	0.18	0.08	0.24	0.02	0.03	0.04
Iron, Fe	mg/L	0.02	0.03	0.01	0.01	0.45	0.01	0.05	0.13	0.01	0.01

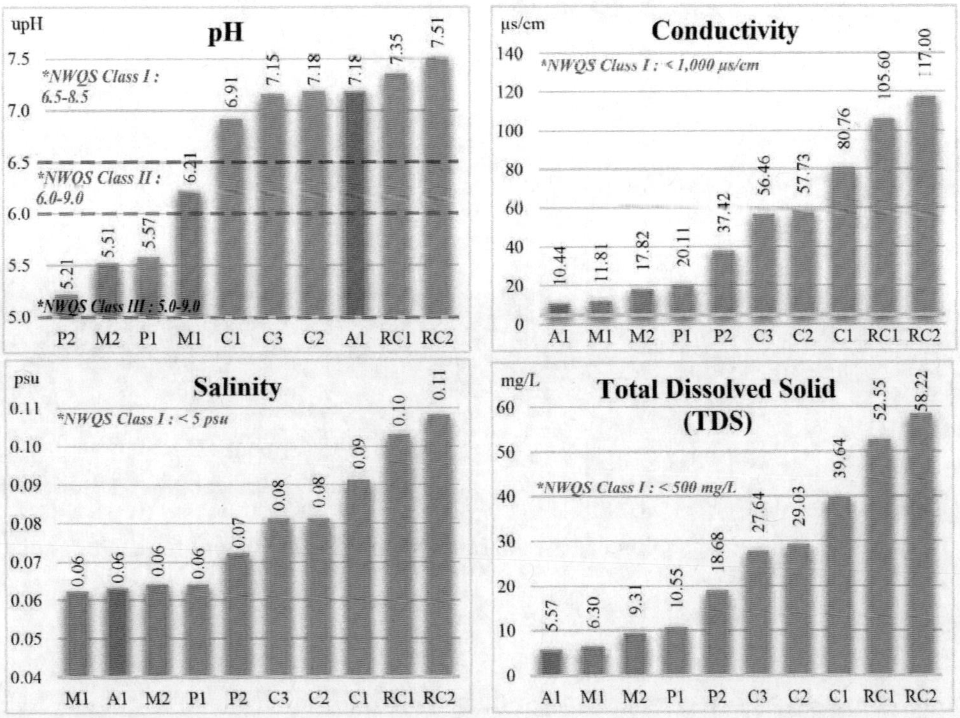

Figure 1. Graphs in ascending order of pH value, conductivity, salinity and TDS for each sample.

IOP Conf. Series: Materials Science and Engineering **401** (2018) 012011 doi:10.1088/1757-899X/401/1/012011

3.1. pH

Clay tile roof-harvested rainwater displays a rather neutral reading, similar to that of ambient rain, indicating that clay tile does not significantly affect runoff pH value. Both polycarbonate roof and metal roof harvest rainwater that shows low pH value, which is in contradict with other similar researches. However the source of acidity is not identified. Lower pH value from metal and polycarbonate roof runoffs caused the high level of aluminium corrosion from the roofs, suggesting that both metal roofs are made up of alloy metals containing aluminium. Although polycarbonate sheets have no metal content, heavy metal contamination can still come from glazing bars and metal crossbraces.

3.2. Conductivity, Salinity and TDS

The graphs indicates the highest level of active ions in rainwater harvested from RC flat roofs, followed by clay tile roofs, polycarbonate roofs and metal roofs. Sources of active ions include dissolved salt or mild acids such as the most commonly found hydrochloric acid. Active ions in the form of calcium or magnesium are also one of the main contributor to the water hardness. In this research, the active ions found in RC flat roof runoffs are mostly from dissolved salt and hardness, given the non-acidic pH value recorded.

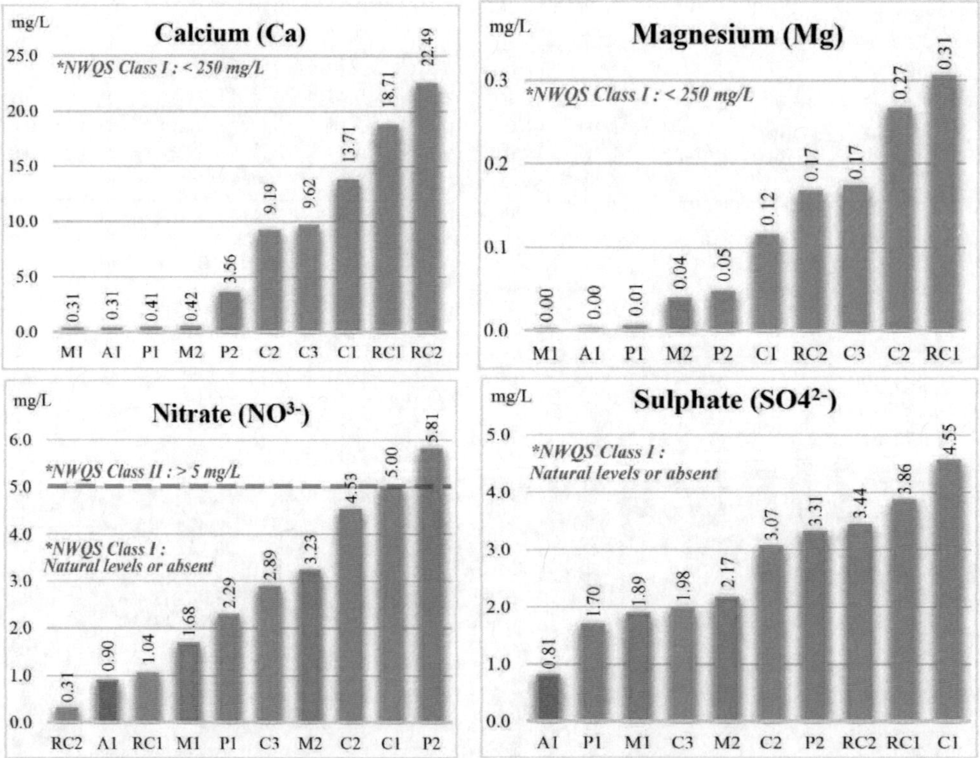

Figure 2. Graphs in ascending order of hardness and mineral content for each sample.

3.3. Hardness

The significant existence of calcium and magnesium content found on clay tile roof and RC flat roof might be due to the dissolution of calcium carbonate and magnesium carbonate compound from the concrete and lime related surfaces - from the screeding or paint.

The higher level of calcium and magnesium found in clay tile roof and RC flat roof runoffs are explanatory to the pH graph as well, due to the alkalinity of both components. Both metal roof and polycarbonate roof yielded relatively soft rainwater quality, requiring less treatment to soften the water for household cleaning purposes.

3.4. *Minerals*

Higher degree of porosity from clay tiles will result in the growth of lichens and mosses, thus showing a higher reading of nitrate contents. [5] Therefore, maintenance for clay tile roof is relatively important due to the surface texture which can easily contain moisture and sedimentations. Nitrate content from metal roof M2 might come from the decaying of fallen leaves or fecal deposited by birds or rodents. The roof was observed to be poorly maintained, resulting in a higher yield of nitrate contamination. C1 showed the highest reading in terms of sulphate pollution due to the low gradient and that the porosity of clay tiles in retaining sulphate related substances.

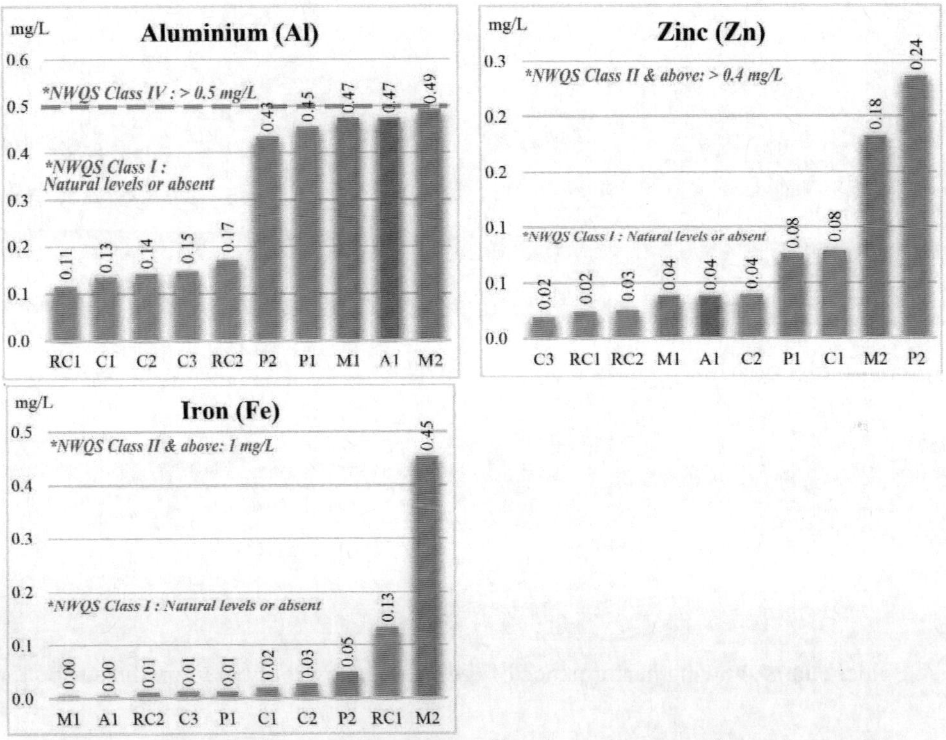

Figure 3. Graphs in ascending order of hard metal content for each sample.

3.5. *Heavy metals*

Given the low pH value in their runoff, P1, P2 and M2 are observed to have rusty surfaces and metal frames, contributing to the corrosion of aluminium compound on the roof surface. M1 can be identified as aluminium coated metal roof or that the component of the metal sheets contained aluminium compound which is easily corroded.

M2 and P2 showed significant reading in the zinc concentration graph. M2 and the metal frames on P2 are observed to be rusty and lack maintenance. The oxidation of alloy metal sheets might result in a mix of corroded metal remaining such as zinc and aluminum on the roof surface, which is then washed down through the runoffs. M2 showed the highest reading for iron content. Runoff from M2 contained high levels of aluminum, zinc and iron, due to the rustiness and debris deposited on the roof from fallen leaves or animal feces.

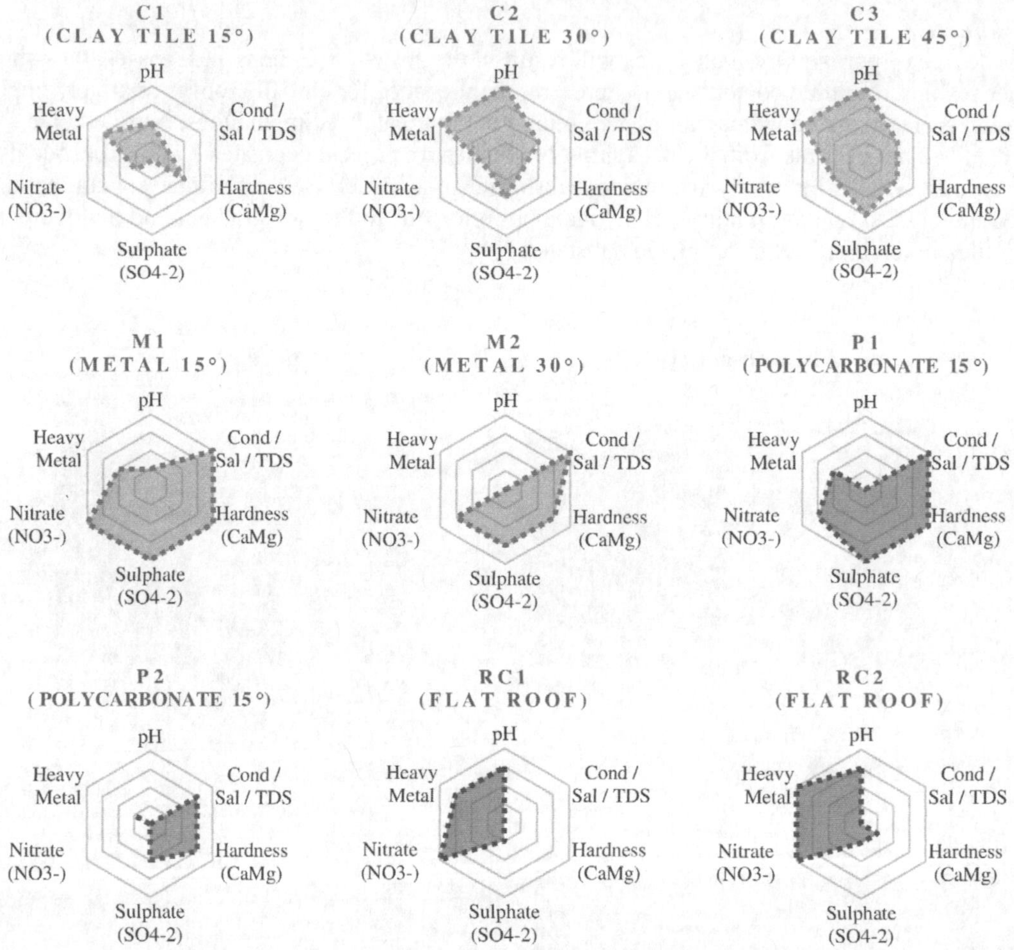

Figure 4. Spider charts showing performance of each case study roof based on the attributes tested.

4. Conclusion

The result observed the best runoff quality from clay tile roof of 45°, metal roof of 15° and polycarbonate roof of 15°; whilst RC flat roofs harvested the most contaminated runoffs. The study shows that all roof runoff quality is not up to potable standard; though can be used for indoor non-potable use with minimal treatment. Besides raising awareness on effects of roof design and material to quality of harvested rainwater, findings of this research are able to identify design criteria of a building roof affecting the roof runoff quality without first flush system. This will aid practitioners in designing sustainable roofs within the local context. This research will serve as a guideline and reference to achieve water and energy efficient roof designs based on the intended usage of roof harvested rainwater, particularly for designers, authorities, manufacturers as well as green rating tools.

This research was conducted with several limitations addressed, including matters related to case study selection and lack of resources. Firstly, the study focused solely on the runoff quality harvested from immediate catchment area – the roof surface, while other RWH mediums such as gutters, rainwater downpipe and storage tanks are not involved. The selection of case studies were done based on observation, where there might be slight variation of recorded roof pitch degree and building height. Proximity between all case studies were also considered, given the lack of human resources during the data collection, which is required to take place within the first flush of rain event. The roofing materials are of different grades and quality, installed at different point of time, with different maintenance routine. Due to the lack of funding for laboratory test, only one rain event is adopted for data collection, thus there is no result comparison done on multiple rain events.

Further research can be done in consideration of the mentioned limitations, where pilot roofs can be installed and observed over a certain period of time to obtain a more accurate data for comparison. The effects of different grades and quality of the same type of roofing material can also be studied to identify the optimal quality desired for respective material with suitable installation method. Research of microbial contaminations on different types of roof material is also important to reduce health related issues from consumption of harvested rainwater and for development of roof materials that are less prone to microbial growth. The study of runoff quality through overall RWH system from collection point to storage can also be useful to optimize the potential of rainwater for domestic use.

5. Acknowledgement

This study was supported by the Centre for Building, Construction & Tropical Architecture (BuCTA), Faculty of Built Environment, University of Malaya. Authors are indebted to Hydrogeology Laboratory, Faculty of Science and Laboratory of Microbial Ecology, Institute of Biological Sciences, University of Malaya as well as all associated members for their generosity in sharing resources and technical assistance throughout the research.

References

[1] Sazakli, E., Alexopoulos, A., Leotsinidis, M., 2007. Rainwater harvesting, quality assessment and utilization in Kefalonia Island, Greece. Water Research 41 (9), 2039-2047.

[2] Fletcher, T.D., Deletic, A., Mitchell, V.G., Hatt, B.E., 2008. Reuse of urban runoff in Australia: a review of recent advances and remaining challenges. Journal of Environmental Quality 37, S116-S127.

[3] Lee, J. Y., Bak, G., & Han, M. (2012). Quality of roof-harvested rainwater – Comparison of different roofing materials. Environmental Pollution, 162, 422-429. doi:10.1016/j.envpol.2011.12.005.

[4] Gobel, P., Dierkes, C., Coldewey, W.G., 2007. Storm water runoff concentration matrix for urban areas. Journal of Contaminant Hydrology 91, 26-42.

[5] Lee, J.Y., Yang, J.S., Han, M., Choi, J., 2010. Comparison of the microbiological and chemical characterization of harvested rainwater and reservoir water as alternative water resources. Science of the Total Environment 408 (4), 896e905.

[6] Forster, J., 1999. Variability of roof runoff quality. Water Science and Technology 39 (5), 137-144.

[7] Olaoye, R. A., & Olaniyan, O. S. (2012). Quality of Rainwater from Different Roof Material. International Journal of Engineering and Technology, 2(8), 1413-1421.

[8] Farreny, R., Morales-Pinzón, T., Guisasola, A., Tayà, C., Rieradevall, J., & Gabarrell, X. (2011). Roof selection for rainwater harvesting: Quantity and quality assessments in Spain. Water Research, 45(10), 3245-3254. doi:10.1016/j.watres.2011.03.036

[9] Despins, C., Farahbakhsh, K., Leidl, C., 2009. Assessment of rainwater quality from rainwater harvesting systems in Ontario, Canada. Journal of Water Supply: Research and Technology-AQUA 58 (2), 117e134.

[10] Simmons, G., Hope, V., Lewis, G., Whitmore, J., Gao, W.Z., 2001. Contamination of potable roof-collected rainwater in Auckland, New Zealand. Water Research 35 (6), 1518-1524.

[11] Chang, M., McBroom, M.W., Scott Beasley, R., 2004. Roofing as a source of nonpoint water pollution. Journal of Environmental Management 73, 307e315.

[12] Abbott, S., Caughley, B., Douwes, J.,(2007) The Microbiological Quality of Roof-collected Rainwater of Private Dwellings in New Zealand. Institute of Food Nutrition and Human Health, Massey University, New Zealand.

IOP Conf. Series: Materials Science and Engineering **401** (2018) 012012 doi:10.1088/1757-899X/401/1/012012

The Usefulness of Main Prayer Hall of Masjid *Agung* Islamic Center of Lhokseumawe, Aceh - Indonesia

Bambang Karsono[1], Zayn Al-Abideen Gregory[1], Dona Rose Amer Koesmeri[1], Awang Hashim bin Awang Sulong[1] and Atta Idrawani Zaini[1]

[1]Reseacher and Lecturer, Department of Architecture, Faculty of Engineering, Universiti Malaysia Sarawak (UNIMAS), Malaysia

kbambang@unimas.my, bkarsono@unimal.ac.id

Abstract. Spatial design of masjids is typically the result of duplicating or replicating typology considered appropriate by the Muslim community. Often, this approach results in a massive centralized main prayer space that is underutilized and may contradict fundamental Islamic principles derived from the Quran and *Hadith* (canonical sayings of Prophet Muhammad ﷺ). This article attempts to reveal the utilization of the main praying hall at Masjid *Agung* Islamic Center (MAIC) of Lhokseumawe. Further, this study challenges the MAIC in light of the Islamic principle of 'avoiding exaggeration'. Observation was conducted over 30 days alongside an exploration of values and principles guiding Islamic Architecture according to Quran and Hadith. Findings show the utilization of the main prayer hall of MAIC averaged 3.24% on a daily basis and 3.26% for *Shalah* (ritual prostration performed five times a day); clearly, a sign of wasted space. Results of this study are expected to stimulate discussion in the area of design and planning of Islamic Architecture toward a wiser design, particularly in Aceh Province of Indonesia.

Keywords: masjid, main prayer hall, avoiding exaggeration, waste of space, simplicity

1. Introduction

Masjid Agung *Islamic Centre* (MAIC) of Lhokseumawe is situated in the center of the city on a land area of 33,748.47m2. Today, the masjid building itself occupies an area of 16,475.80m2. Inside, there is a main space with an area of 3,558.33m2 dedicated for conducting the prayer. Other rooms are provided, such as a secretariat, imam's room, meeting hall, ablution area and office units. Administratively, MAIC is located in Banda Sakti district of Lhokseumawe. The district covers 112.40 hectares with a population of 80,769 people. The population is spread over 18 gampongs (villages) with a gender ratio of 100 women to 99.66 men. Other than MAIC, there are 13 other masjids spread across the gampongs in Banda Sakti district. For an approximation, if an area of 0.72m2 (0.6x1.2m) is used for performing shalah, then the main prayer hall of MAIC is able to accommodate 4,942 people for that activity.

Many hadiths explain the numerous ways masjids were used during the life of Prophet Muhammad ﷺ, particularly the Masjid al Haram in Mecca and Masjid al Nabawi in Medina. The Hadith literature, describing the behaviour and attitude of Prophet Muhammad ,ﷺ constitute a normative reference for the individual Muslim in all aspects of life, including when designing and using space within a masjid. Narratives of the social life around the Kaaba, which is now the center of the Masjid al Haram, illustrate

Content from this work may be used under the terms of the Creative Commons Attribution 3.0 licence. Any further distribution of this work must maintain attribution to the author(s) and the title of the work, journal citation and DOI.
Published under licence by IOP Publishing Ltd

that the masjid was utilized for congregation for all levels of society. Political and diplomatic activities were conducted nearby. Prophet Muhammad ﷺ held many discussions and preached within this space.

Such hadith show the sacred space around the Kaaba was not merely a location for prayer but for a variety of other communal activities. Masjid al Nabawi was utilized the same way too as a center for social, political and worship activities. Other Hadiths provide support for the use of a masjid as a shelter, an educational institution, health facility, and a site for celebration and recreation activities nearby.

Figure 1: Illustration of Masjid al Nabawi during Prophet Muhammad's ﷺ time, with its various activities in the spirit of Islam.

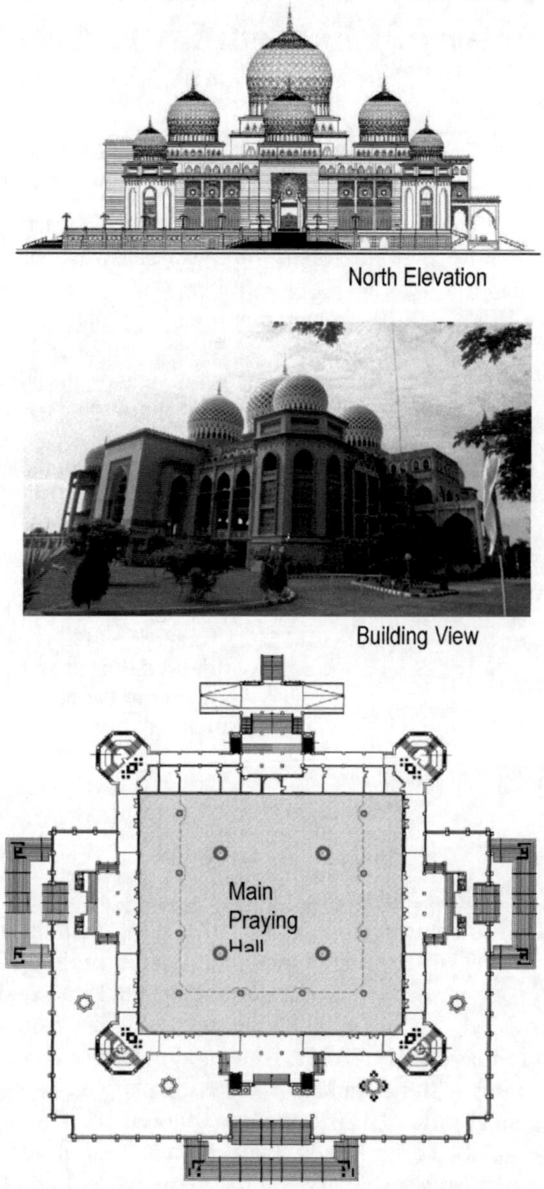

Figure 2: MAIC of Lhokseumawe

Architecturally, spaces within Masjid al Nabawi had a very flexible arrangement allowing for multifunctional spaces. The spaces had full utility without any wasted space. A spacious courtyard surrounded by walls made the area safe and comfortable enough to allow for congregational shalah training for physical activities, sports activities, gatherings, discussions and educational activities to take place. Also, the roofed spaces were flexibly used for various activities such as congregational shalah, study and discussion. Of course, the existence of a large courtyard for congregational prayer space was suitable for the dry climate of Medina and the entire Arab peninsula. Prophet Muhammad ﷺ reminded his followers not to exaggerate in building a masjid as stated in the following hadith:

Ibnu 'Abbas relates that Prophet Muhammad ﷺ *said:*
"I was not ordered to build high masjids."
(Narrated by Abu Dawud)

Anas ibn Malik relates that Prophet Muhammad ﷺ *said:*
"Judgement Day will not arrive until humans exaggerated in building masjids."
(Narrated by Ahmad, Abu Daud, An-Nasa'i and Ibnu Majah)

'High masjids' in the hadith above literally implies exaggeration in space planning and construction. Therefore, a masjid can be understood to be not a monumental structure that is utilized simply for the purpose of prayer alone. Rather, a masjid is intended to be a place for Muslims to interact and conduct various activities that provide benefit to the religious community.

Prophet Muhammad ﷺ forbade his *ummah* (religious community) from exaggeration or excess including in building and selecting material for a building. Islam teaches a simple, humble and functional architectural language as described within the following hadiths:

> *"Anas bin Malik said: Prophet Muhammad ﷺ one day saw a massive building with dome on top it and then said: What was it? Prophet Muhammad answered: it was a building belong to someone..., one of the Ansar. Prophet Muhammad ﷺ did not mention any single word thus raised a big question among prophets' companions. n its owner greeted, Prophet Muhammad ﷺ turned away his face and step away from him. The owner repeated his greet several times and Prophet Muhammad ﷺ reacted in the same way, thus the owner realized Prophet Muhammad's ﷺ anger toward him. Finally, he asked about his problem to other companions and said: I swear to Allah, that I did not understand Prophet Muhammad ﷺ attitude toward me. The companions answered the Prophet Muhammad ﷺ behave in that way after He saw a massive building with the dome belong, to you. He then returned home and destroyed the building to the ground. One day Prophet Muhammad ﷺ saw to the direction of the dome building and He did not discover the dome building anymore. The Prophet Muhammad ﷺ asked: What had happened to the dome building? The Companions answered Him: "the owner had complained about your attitude toward him that you had turned your face away from him when you met him and when we told him the cause of it then he destroyed his dome building. Prophet Muhammad ﷺ said: "Every building was a slander to his owner unless human could not live his life without it". (Sunnah Abu Dawud, Vol. III, pp 1444-1445)*

The usefulness of space is a fundamental value in architecture. Uselessness can cause waste and disadvantages toward certain matters, especially if it is done repeatedly and exaggeratedly. In the Islamic worldview, knowledge and understanding toward the rule of usefulness and avoidance of disadvantages will benefit to obtain *maslahat* and elude *mafsadat* in the practice of architecture. The explanation of the previous hadiths has showed the spirit toward simplicity and humbleness for the true beauty according to Islamic principles.

2. Methods

This study used a normative case study method aimed at developing knowledge intensively about an object (Zeisel, 1980). This research was conducted by surveying the hadith literature which then were used as a framework to explore the usefulness of the main prayer hall of a Masjid. Data was obtained from dimension mapping of the main prayer hall of the MAIC, observing and measuring the space used during the 5 daily shalah and for *Jum'ah* (Friday sermon). Observation and measurement was conducted within 30 days. The observed data of prayer attendance were recorded and tabulated with the hall measurements to yield the utilization of the area, compiled and displayed using *Microsoft Excel* in the form of tables and diagrams. Analysis and discussion were performed through a thorough observation of phenomena of usefulness of the main praying hall.

3. Results and Discussion

Result are plotted consecutively through the tables that follow. Table 1 tabulated average percentage values (%) of used space according to day and time of shalah. The table showed the highest average utilization value happened on Friday during Jum'ah at 31.26%, equivalent to 1,112.33m2 of space used, leaving 2,445.99m2 unused. Meanwhile, the lowest average utilization value occurred on Sunday during *Isya* (night) prayer at 0.63%, equivalent to 22.42m2 of area used, 3,535.91m2 vacant.

Table 2 showed tabulated average data (%) of used area according to prayer time. The highest average utilization occurred at *Dzuhur* (noon) prayer at 5.48% (including Jum'ah, which replaces Dzuhur on Friday), equivalent to an area of 194.99m2 of space. By contrast, the lowest average utilization happened at *Shubuh* (dawn) prayer at 1.10% or 39.14m2 of space used. In total, the average utilization value for 30 days was 3.26%, equivalent to 116.00m2 of space. These findings indicate a massive amount of vacant space on average, reaching 96.74% of space.

Table 1. Average (%) of utilization of main prayer space according to day and time of shalah, 30 observation days

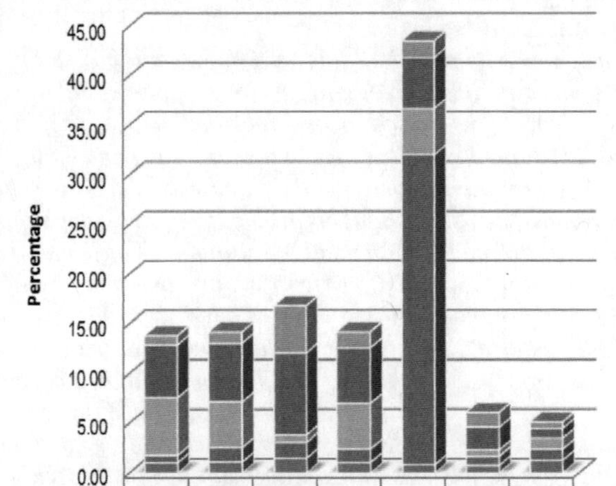

	Mon	Tue	Wed	Thu	Fri	Sat	Sun
Isya	0.91	1.13	4.73	1.58	1.68	1.48	0.63
Maghrib	5.30	5.89	8.28	5.58	5.16	2.29	1.01
Ashar	5.85	4.63	0.77	4.65	4.67	0.65	1.09
Dzuhur	0.73	1.44	1.58	1.34	31.26	0.91	1.11
Subuh	1.05	1.13	1.50	1.07	0.85	0.77	1.29

Table 2. Average value (%) of utilization according to Shalah, 30 days' observation

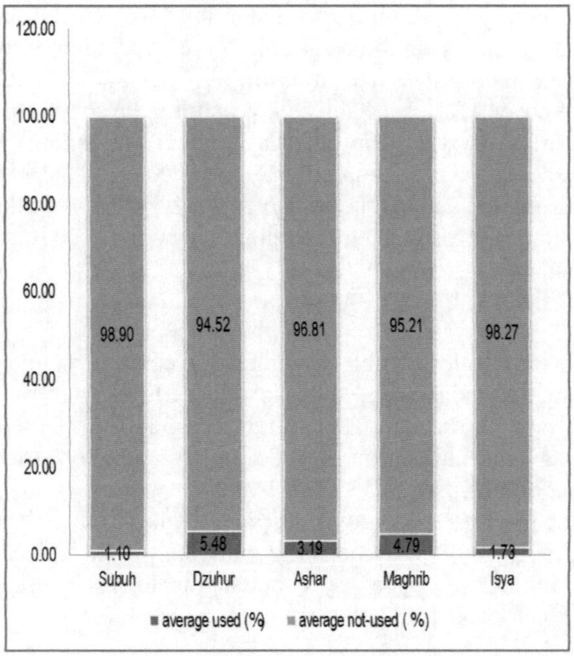

In Table 3 average utilization is according to day of the week. The highest average utilization occurred on Friday at 8.72%, equivalent to 310.28m2. The lowest average utilization happened on Sunday at 1.03%, equivalent to 36.65m2 of used space. In total, the average value of utilization according to praying time for 30 days was 3.24%, equivalent to 115.28m2 area coverage. A massive 96.76% of hall space remained unused.

Results show the excessive amount of area unused at the main prayer hall of MAIC with 95% of unused area. This must be considered excessive. In building of a masjid, increasing cost for construction is often the most important issue that was seriously discussed; as a consequence, often the planned budget for construction would be categorized as excessive.

High construction cost with only less than 4% area properly utilized is one of the strongest signs of excess. Furthermore, appropriate analysis for the main prayer space of a masjid relative to population and situation of a specific community was clearly not carried out effectively.

On the other hand, the massive size of the prayer hall requires high energy for maintenance and routine use of electricity which in turn contributes to excessive operational cost. High operational costs become a waste where less than 4% of the space itself is actively used.

This situation is in high contrast to the Islamic way of life practiced by Prophet Muhammad ﷺ as shown in the following hadith:

Narrated by Muhammad bin Ziyad: I had heard Abu Hurairah said,
"When they saw a man, the important person of Bahrain, swipe the ground using their clothes, I said, 'the important person is coming, the important person is coming.'"
Prophet Muhammad ﷺ said, "Indeed, Allah Most High would not bless those who swipe the ground boastfully (Shahih Muslim. Hadith No 1359. pp. 772.)

IOP Conf. Series: Materials Science and Engineering **401** (2018) 012012 doi:10.1088/1757-899X/401/1/012012

Table 3. Average value (%) of utilization of main prayer hall according to day of the week, 30 observation days.

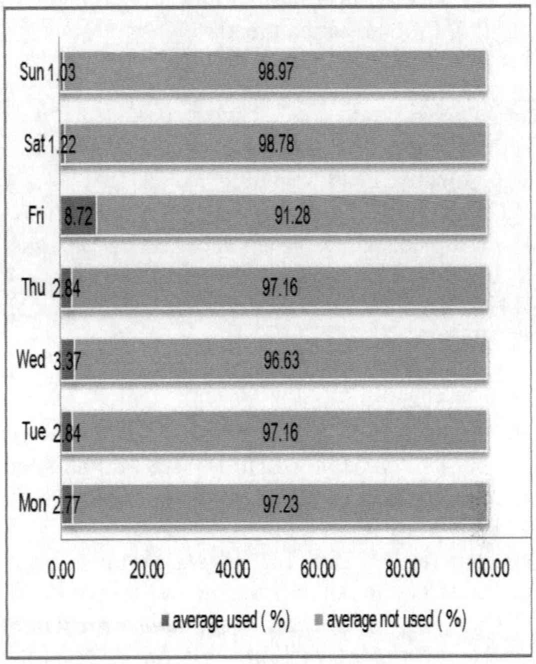

Narrated by Anas bin Malik:
When we sat with Prophet Muhammad ﷺ inside the masjid, someone came riding a camel. He ordered his camel to kneel at the masjid yard, tied its front feet and asked, "Who among you is called Muhammad?" At the time Prophet Muhammad ﷺ was sitting among companions leaning on both of his hands. We answered, "The white man sitting leaning on his hand." The man looked toward Prophet Muhammad ﷺ, "Oh the son of Abdul Muththalib." Prophet Muhammad ﷺ said, "I came here to answer your questions."The man asked Prophet Muhammad ﷺ, "I wanted to ask something that seems hard for you to answer. So do not be angry." Prophet Muhammad ﷺ said, "Ask anything you want to ask." The man said, "I ask in the name of your God, the God of people before you, did Allah Most High send you as a prophet for entire humankind?" Prophet Muhammad ﷺ answered, "For the sake of Allah Most High, that is right." Then, the man asked, "In the name of Allah Most High I ask you. Did Allah Most High order you to conduct shalah five times within one day and night?" Prophet Muhammad ﷺ answered, "In the name of Allah, that is right." Moreover, the man asked, "In the name of Allah! Did Allah Most High order you to fast during the month of Ramadhan?" Prophet Muhammad ﷺanswered, "Right, in the name of Allah Most High." Further the man asked, "In the name of Allah Most High! Did Allah Most High order you to take charity from the wealthy and distribute it to the poor?" Prophet Muhammad ﷺanswered, "In the name of Allah, that is right." Then the man said, "I confirm my faith in what He has sent down to you. I was sent by my tribe as a messenger, my name is fulan from the family of Bani Sa'd bin Bakar." (Shahih Muslim. Hadits No-1387. pp. 784.)

The hadith above illustrates how the Prophet Muhammad ﷺ conducted himself among his companions inside the masjid. He did not show himself as an important person as a messenger of Allah Most High; rather, Prophet Muhammad ﷺ carried himself in an egalitarian manner. His demonstration of simple and humble living is an important value that informs every aspect of the believer's daily life. This is highly relevant when it comes to building of a masjid which is the architectural manifestation of the spiritual values of the *ummah*.

In an effort to reduce excess and practice simplicity when building a masjid, main prayer spaces should be designed appropriately to avoid potentially useless space. Functionally, the main prayer hall of MAIC could avoid exaggeration by arranging two groups of spaces, a fixed praying space and a flexible area for various relevant activities to take place. The fixed space would provide space appropriate for conducting shalah (specific to MAIC: ± 3.26% x 3,558.33 = 116.00 m2). The remaining area can be a flexible space for other relevant activities as has been exemplified by Prophet Muhammad ﷺ. Also, for temporary praying activities that require larger space such as Jum'ah and *Eid* (holiday) prayers, it can take the advantage of the flexible space to accommodate exceptionally large congregations.

4. Conclusion
Findings indicated that the main prayer hall must be considered an exaggerated and wasteful space. Observations over 30 days revealed that the usefulness of space is less than 4% (3.24% according to day of the week and 3.26% according to praying time) out of the total available main prayer space.

IOP Conf. Series: Materials Science and Engineering **401** (2018) 012012 doi:10.1088/1757-899X/401/1/012012

The waste of space can be avoided by mature design considering the capacity from the specific population of an area and the expected number of community members using the space. With that, the useless space within masjids can be avoided as an implementation of simplicity and humbleness as taught by Prophet Muhammad ﷺ.

References

[1] Al Faruqi, I. & Al Faruqi, L.L., *The Cultural Atlas of Islam,* New York: Macmillan Publishing Company, 1986

[2] Creswell, KAC., *A Short Account of Early Muslim Architecture*. Beirut: Librarie du Liban, 1968

[3] Frishman, M. & Khan, H.U., *The Mosque History, Architectural Development and Diversity*, London: Thames & Hudson Ltd., 1994.

[4] Hamka, *Sejarah Perkembangan Pemurnian Ajaran Islam di Indonesia*, Jakarta: Tintamas Djakarta., 1958.

[5] Michell, G. Ed., *Architecture of the Islamic World: Its History and Social Mean*ing, London: Thames & Hudson Ltd., 1978.

[6] Mohd. Tajuddin M Rasdi, *Crisis in Islamic Architecture. KALAM Papers: Crisis in Islamic Architecture*, Johor Bahru: Pusat Kajian Alam Bina Dunia Melayu (KALAM), 2003.

[7] Utaberta, N., *Arsitektur Islam: Pemikiran, Diskusi dan Pencarian Bentuk*, Yogyakarta: Gadjah Mada University Press, 2008

[8] Utaberta, N., *KALAM Papers: Makna dan Arti Keindahan dalam Arsitektur Islam.* Johor Bahru: Pusat Kajian Alam Bina Dunia Melayu (KALAM), 2003

[9] Utaberta, N., *Muslim Architecture in Peninsular Malaysia: Clasification of Styles and Probable Socio-Political Influence of Mosques.* Johor Bahru: Pusat Kajian Alam Bina Dunia Melayu (KALAM). 2003

[10] Zein, A.B., *Masjid-masjid Bersejarah di Indonesia*, Jakarta: Gema Insani Press., 1999.

[11] Ziesel, J., *Inquiry by design: Tools for environment-behaviour research*. California: Brooks/Cole

Functional Attachment in Kuching Riverfront Promenade, Sarawak - Malaysia

Bambang Karsono[1], Nurakmal Abdullah @ Goh Tuo Ho[1], Dona Rose Amer Koesmeri[1], Zayn Al-Abideen Gregory[1], Yon Syafni Samat[1]

[1]Reseacher and Lecturer, Department of Architecture, Faculty of Engineering, Universiti Malaysia Sarawak (UNIMAS), Malaysia

kbambang@unimas.my, bkarsono@unimal.ac.id

Abstract. The relationship between people and certain places can stimulate the sense of place, this process is recognized as place attachment. Place attachment also indicates the relationship between physical image and its function, through this process the sense of a place can be established. These concepts focus on the physical elements and activities together with the user-influenced meanings that will create the identity and the meaning of the place. Allowing for this issue, the research examines the functional attachment which will affect the level of attachment in Kuching Riverfront Promenade (KRP), a popular public space among local people. By using indirect method to simplify common patterns and human-specific patterns of the place, this research covers subject of place attachment. Questionnaire survey was carried out at several areas at the promenade with an overall of 165 respondents and 18 stall operators were interviewed. The findings point out that the respondents have strong association with the local environment and designate the importance of place as economics dependence and recreational place. The level of functional attachment to KRP is between level 1 to level 4. Measuring at level 4 involves a higher commitment in loyalty, goals and responsibility to the environment.

Keywords: *place attachment, functional attachment, sense of place*

1. Introduction

A place cannot be understood without taking into account the three components i.e. physical features or forms; activities and functions; and meanings or symbols [1]. It explains that a place consists of physical form, activity and meaning. Canter [2] has similar opinion that a place is a combination of three aspects consisting of the activities, the concept of place and the physical environment. The strong relationship that exists between emotional (affective) and functional in such space or relationship between people and certain places can stimulate the sense of place, this process is recognized as place attachment [3]. Place attachment in this study indicates the relationship between physical image and its function or relationship between the user and a particular place. Through this process the meaning of a place can be developed. This research argues that the consequences of inappropriate physical change may affect the meaning of the place and the attachment of the place will be gradually reduced. The research investigates this issue by using place-based concepts and principles. These concepts and principles focus on the physical elements and activities together with the meanings possessed by the users that will generate the identity of the city and the meaning of the place. Within this framework, the study focuses on the dimension of place attachment in examining the psychological aspects of the place and its relation to

Content from this work may be used under the terms of the Creative Commons Attribution 3.0 licence. Any further distribution of this work must maintain attribution to the author(s) and the title of the work, journal citation and DOI.
Published under licence by IOP Publishing Ltd

the physical component. Therefore, this research is intended to identify the place attachment of the riverfront promenade in a city that is popular, in Kuching, Sarawak, Malaysia.

Kuching Riverfront Promenade (KRP) was chosen as a case study. It is a popular public open space located in front of the main road of Kuching, a city in the state of Sarawak, Malaysia. KRP was officially opened to public on 3rd September 1993. The development of KRP has not only introduced public open space, but also a recreational place, which generated the development of a city landmark. Furthermore, KRP which is commonly known as "the People Place", is furnished with facilities for entertainment, refreshment, relaxation, cultural and art performances [4]. KRP has received three international awards in 1994. This study focuses to recognize the exact form and level of functional and emotional attachment of the user in KRP. This research is important in identifying the localities of a place within a city as a place character which is meaningful to the people.

Figure 1: Kuching Riverfront Promenade (KRP)

2. Literature Review

Place attachment is the main concept in this study. Definitions of place attachment consist of positive emotional attachments between individuals within their setting and environment [5]. The psychological situation of well-being which is generated from access into places or circumstances will also affect human perceptions about the place [6]. Meanwhile Hummon [7] argues that the level of human emotional investment within the place indicates the level of successful neighbourhood to become a place. Place attachment and related concepts of sense of place have overwhelmed the theory for the past three decades [1] [2] [5]. In the last 15 years, various studies have been made to explore these concepts empirically. This empirical study focusses on measuring the intensity of attachment.

Place attachment arises through the influence of relationships between humans and places because of satisfaction, value and the experience of people to a special place [8]. The quality of a place has been measured to the way how it is perceived and valued by people. It is very important that attention should be given to the influence of place dimensions, since urban environment purposes are to be utilized by society and how they perceive it [9]. Place can be a success if it is constructed based on the perception of the users, the question whether a place has a such meaning or not for people can be explained through understanding the attachment of the place in depth.

Recent research explains the sense of place and place attachment by using quantitative method. There are several variables used to measure attachments include: familiarity, period of discourse, reliance, place of identity, sense of belonging and sense of place [1] [10] [11] [12]. Shamai [11] develops seven levels of measurement criteria as a tool to understanding the sense of place or sense to different size of territories level (refer to Table 1). According to him, not all ratings are suitable for use as all types of places. Based on measurements criteria, the level of attachment can start from the 0 (zero) to 6th level. In this research, measurements of sense of place can help researchers in identifying the form and level of functional attachment in KRP.

Table 1: Sense of place level, according to Shamai [11]

Level	Sense of place criteria	Keywords
0	Has no sense of place	There is no attachment
1	Knowing being in place	Knowing; understanding symbols
2	Has sense of belonging to the place	Sense of belonging; togetherness; sense of same fate; respected symbols
3	Has attachment to the place	Emotion; meaning; experience; symbol; identity; personality; uniqueness; difference
4	Understand the place goals	Share goals; loyalty; obedience
5	Has a place engagement	Commitment; engagement; behaviour; investment of human resources in group activities; actively involved.
6	Sacrifice for the place	Sacrifice; good commitment

3. Methodology

The basic dimensions of place attachment are known as a perception of place as something unique or different and assessment or meaning whether it is negative or positive [13] To assess perception of the environment, scholars generally use two methods, i.e. direct and indirect assessment [14] [15]. Indirect methods include self-reports (questionnaires, interview, checklist and description) and time-sampling [16]. The direct method involves the phenomenology approach, to check and describe the human situation and the events that occur, as happens spontaneously in everyday life [17]. This approach can be generated through a critical description of what has been happened, having thrown away all the initial hypothesis about the object [18]. The goal is to understand qualitatively the uniqueness of the holistic meaning of a place.

This research used indirect methods (questionnaires, interviews, observation) to clarify common patterns and human-specific patterns of the place. The survey was conducted through 165 respondents that consisted of the mobile users (82) and static users (83). The static users are the crucial users such as the shop-owners, the shop-keepers and sidewalk vendors while the mobile users are the visitors, students and local residents who came to visit the place. At the same time eighteen (18) in depth interviews were also carried out in the study area. The respondents represent the main users of the place such as: shop owners, shop keepers, sidewalk vendors, visitors, students and locals. The age of the respondents is mainly between 18-24 and 25-49 years old, where nearly 60% of them are women. Indicator for each attribute constructed in the questionnaire for the evaluation is cross examining with the literature. The format of this assessment is planned based on a 5-point scale derived from the fairness quality measurement. In the analysis, the result of the evaluation is obtained through triangulation of the data, interview and observation. It is assumed that the nature of attachment is determined by the place attributes with the function of riverfront promenade.

4. Result and Discussion

4.1. Form of Functional Attachment

The response from the interview as shown in Table 2 suggests that the attachment has been expressed in terms of length of engagement, level of familiarity, level of dependency, level of satisfaction and sense of comfort. This shows the importance of open space in supporting economic and social activities. There is also indication that a place relies on earning fixed income and business opportunities during events or festivals. Level of satisfaction is expressed through the availability and diversity of public facilities, street access, the intensity of visitors and buyers. Physical and environmental comfort is reflected in the ease of accessibility, facilities associated with weather protection. Respondents were generally satisfied with the physical enhancement of open space and path, various and diversity of products, and attract visitors and buyers. While other respondents mentioned the significance of location and pedestrian intensity in supporting their attachment to KRP

Table 2: Functional attachment indicators (result from in-depth interview)

Indicator	Place
	Kuching Riverfront Promenade (KRP)
Engagement	– Engagement from morning to evening. – Attachment to the place is due to long periods of engagement. – Attachment to the place because of long-term relationship. – Continuous visit especially during holidays and new year increases familiarity. – Street vendors and trading activities – Leisure and entertainment – The attraction for food and drinks
	Keywords: ***Familiarity, Street Vendor, Social, Relationships.***
Familiarity	– Very familiar with the place. – Can describe the area very well, very imageable – Can illustrate physical changes in detail – Know most other hawkers – Always engage with the street vendors that are visited frequently and very familiar with them
	Keywords: ***Images, Changes, Hawkers, Events.***
Dependency	– Place to acquire income – The place is chosen because of its economic potential and the attachment to the place. – A good understanding over the space due to its length of engagement. – Attachment to the place because it is easy to earn money at. – Attachment is due to business profits
	Keywords: ***Income, Economy, Attachment, Control***
Satisfaction	– Satisfied with various transportation and convenience facilities – The ideal place to maintain a living and daily necessities – Satisfied with the various items of food and drinks offered at a low price – Satisfied with the place as it attracted many people to visit – A special highlight of food and drinks
	Keywords: ***Diversity, Sustainable life, Low price***
Comfort	– A comfortable place to work and visit because of its convenience facilities and cheaper price – Happy and comfortable with the place – Easy to get places to eat, comfort access and walking
	Keywords: ***Facility, Price, Access, Walk***

4.2. Level of Functional Attachment

Result from questionnaire survey on the functional attachment, generally indicates a more positive answer. Respondents were asked to respond to statements showing their attachment to the place. The measurement value of 2.72 out of 4.0 (as summarized in Table 3) expresses higher than the average value. This indicates that the respondents have strong functional attachment to KRP. Respondents strongly agree that KRP is the best place for their purpose, they feel comfortable being there as compared to being at other open places in the city and are satisfied with the improvements that have been made to the area. The response to the statement 01 (No other location is better than this place) has very strong functional attachment (2.88). Results show that (shown in Figure 2) more positive responses were collected from respondents at KRP. They felt that open spaces and street space satisfy them and function appropriately. They also felt very comfortable being there compared to other places in the city.

Table 3: The level of functional attachment to KRP based on the average value

Component	Statement		KRP	
			Mean	*Std. Dev*
FUNCTIONAL ATTACHMENT N=165	01	No other location is better than this place (Best)	2.88	*.665*
	02	I feel more comfortable in this location (comfortable)	2.77	*.682*
	03	This place is the best location to meet my goals (Goals)	2.68	*.680*
	04	This location is very important to me (Important)	2.53	*.691*
Response format: 1= strongly disagree, 4= strongly agree		Average Value	**2.72**	

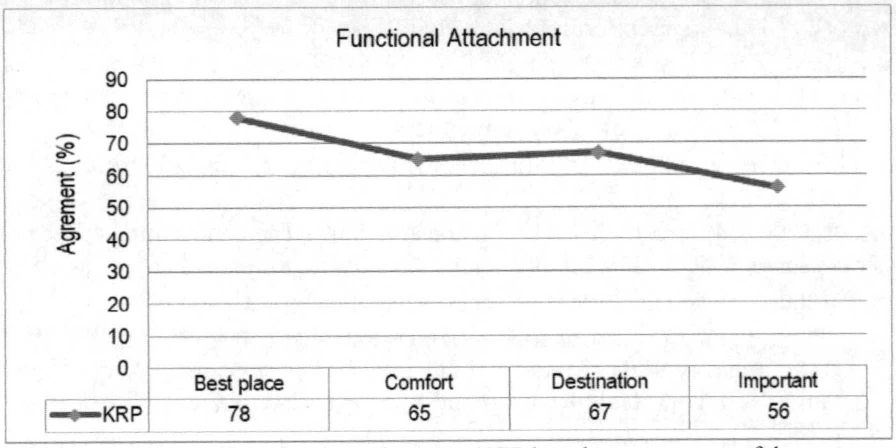

Figure 2: Level of functional attachment to KRP based on percentage of the response (N=165)

5. Conclusion

Based on the result, it is concluded that the respondents (static and mobile) described the importance of place as economic dependence place and recreational place. In short, the respondents felt that KRP is a suitable place because of its functional access between open space and the road. They also suggested that KRP is the best option in supporting their activities compared to other areas. This area is also very important in their lives as it becomes their place of dependence for personal needs and a major place of livelihood. It is concluded that the attributes associated with the dominant function in KRP are influenced by the form of functional attachment. According to the measurement of the sense of place developed by Shamai [11], it can be concluded that the level of functional attachment to KRP is between level 1 to level 4. Measuring at level 4 involves a higher commitment to the environment which is indicated by the loyalty to daily economic activities, loyalty to business and society, and loyalty to the place as a place to earn income and daily necessities. The results also demonstrated that there are some slight different responses between static and mobile respondents. Static respondents expressed their sense of identity and place dependency, while mobile respondents were attached to the role of function of open space in KRP. The results of this study support the assumption that the level of functional attachment will vary according to the user's role in the place.

References

[1] Relph, E. (1976). *Place and Placelessness*. London: Pion Publication.
[2] Canter, D. (1977). *The psychology of place*. London: The Architectural Press Ltd.
[3] Karsono, B., Deni, Fithri, C.A., (2016), Assessment of Functional and Emotional Attachment in Malacca Riverfront Promenade. *Jurnal Teknologi (Sciences & Engineering)* 78:5 (2016) 153–157.

[4] Karsono, B., Indira, S.S., Deni (2016), The Significance of Uniqueness, Comfort, Security and Safety to Place Attachment, *Jurnal Teknologi (Sciences & Engineering)* 78:5 (2015) 179–183

[5] Altman, I., & Low, S. (Eds.). (1992). *Place Attachment*. New York: Plenum Press.

[6] Giuliani, M. V. (2003). Theory of Attachment and Place Attachment. In M. Bonnes, T. Lee & M. Bonaiuto (Eds.), *Psychological Theories for Environmental Issues*. Adershort, UK: Ashgate Publishing Company.

[7] Hummon, D. M. (1992). Community Attachment: Local Sentiment and Sense of Place. In I. Altman & S. Low (Eds.), *Place Attachment*. New York: Plenum Press.

[8] Bonnes, M., & Secchiaroli, G. (1995). *Environmental Psychology: A Psychosocial Introduction*. London: Sage Publications.

[9] Carmona, M., Heath, T., Oc, T., & Tiesdell, S. (2003). *Public Places - Urban Spaces, The Dimensions of Urban Design*. Oxford: Architectural Press.

[10] Proshansky, H. M., Fabian, A. K., & Kaminoff, R. (1995). Place Identity: Physical World Socialisation of the Self, Giving Place Meaning in L. Groat (Ed.), *Readings in Environmental Psychology* (pp. 87-113). San Diego: Academic Press.

[11] Shamai, S. (1991). Sense of Place: An Empirical Measurement. *Journal Geoforum, 22*(3), 347-358.

[12] Stedman, C. R. (2003). Is it really just a social construction? : The contribution of the physical environment to sense of place. *Journal of Society and Natural Resources, 16*, 671-685.

[13] Gustafson, P. (2001). Meanings of Place: Everyday experience and Theoretical Conceptualizations. *Journal of Environmental Psychology, 21*, 5-16.

[14] Proshansky, H. M., Ittelson, W. H., & Rivlin, L. G. (1976). *Environmental Psychology: People and Their Settings*. New York: Holt, Rinehart and Winston.

[15] Gifford, R. (1987). *Environmental Psychology: Principles and Practice*. Boston: Allyn and Bacon, Inc.

[16] Lynch, K. (1960). *The Image of the City*. Massachusetts: MIT Press.

[17] Von Eckartsberg, R. (1998). Introducing Existential-phenomenological Psychology. In R. Valle (Ed.), *Phenomenological inquiry in psychology* (pp. 3-20). New York: Plenum.

[18] Seamon, D. (1982). The Phenomenological Contribution to Environmental Psychology. *Journal of Environmental Psychology, 2*, 119 - 140.

IOP Conf. Series: Materials Science and Engineering **401** (2018) 012014 doi:10.1088/1757-899X/401/1/012014

Creating Multiple Seasons Spatial Model (Maps) to Improve and Habilitate the Marshes Area in Southern Iraq

Ameera Meki Jassim[1], Mohd Sanusi S. Ahamad[2]

[1, 2] School of Civil Engineering, Engineering Campus, Universiti Sains Malaysia, 14300 Nibong Tebal, Pulau Pinang, Malaysia
[1] Corresponding author: amj12_civ017@student.usm.my , ameera_meki@yahoo.com
[2]cesanusi@usm.my

Abstract: Climate factors are one of the main criterion in the study of any area especially for sustainable development goals. The year in Iraq has four seasons. The important climate factors for the marshes area in Southern Iraq are the temperature, the rainfall, the wind speed and the humidity. This research explained the composition (creating) of 24 climate maps representing the dynamic climate factors for the four seasons experienced by the marshes area in Southern Iraq that is located in three governorates, i.e. Missan, Thi-Qar and Basra. The methodology applies the spatial Interpolation Method using the GIS - IDRISI software. The climate maps of a particular area can be used for multi - purpose study either for development or for preparing a project such as ecotourism. Their capability in dynamic mapping and analysis are one of the most useful applications of GIS for sustainable environmental planning and management. The GIS spatial analytical technique applied in this study can reduce time and cost by 20 – 30 % compared to traditional methods. The created dynamic climate maps for each required climate factor are in high resolution (36m x 32m) pixel. In addition, this method of obtaining climate factors values is the ideal and one of few methods in difficult areas where there are mountains, forests…etc. or in dangerous areas which have such as predators , remnants of war…etc.

Keywords: Interpolation method, spatial climate model (maps), Marshes Area Southern Iraq, Mesopotamian Marshes, climate factors.

1. Introduction

The weather is basically the way the atmosphere is behaving, mainly with respect to its effects upon life and human activities. The difference between weather and climates is that weather consists of the short - term (minutes to months) changes in the atmosphere. Most people think of weather in terms of temperature, humidity, precipitation, cloudiness, brightness, visibility, wind, and atmospheric pressure, as in high and low pressure. In most places, weather can change from minute - to - minute, hour - to - hour, day - to - day, and season - to - season.

Climate however, is the average of weather over time and space. An easy way to remember the difference is that climate is what you expect, like a very hot summer, and weather is what you get, like a hot day. In short, climate is the description of the long - term pattern of weather in a particular area. Some scientists

Content from this work may be used under the terms of the Creative Commons Attribution 3.0 licence. Any further distribution of this work must maintain attribution to the author(s) and the title of the work, journal citation and DOI.
Published under licence by IOP Publishing Ltd

IOP Conf. Series: Materials Science and Engineering **401** (2018) 012014 doi:10.1088/1757-899X/401/1/012014

define climate as the average weather for a particular region and time period, usually taken over 30 years. It's really an average pattern of weather for a particular region. When scientists talk about climate, they're looking at averages of precipitation, temperature, humidity, sunshine, wind velocity, phenomena such as fog, frost, and hail storms, and other measures of the weather that occur over a long period in a particular place. When we talk about climate change, we talk about changes in long - term averages of daily weather.

The important reason for studying climate and climate changing is that it will affect people around the world. Rising global temperatures are expected to raise sea levels, and change precipitation and other local climate conditions. Changing regional climate could alter forests, crop yields, and water supplies. It could also affect human health, animals, and many types of ecosystems. Deserts may expand into existing rangelands, and features of some of our National Parks and National Forests may be permanently altered. Human activities have altered the chemical composition of the atmosphere through the buildup of greenhouse gases primarily carbon dioxide, methane, and nitrous oxide (NASA TV 2005).

Geographic information system (GIS) can offer significant support in creating multiple seasons' climate maps. Their capability in dynamic mapping and analysis are one of the most useful applications of GIS for sustainable environmental planning and management. This paper attempt is to show the significance of the composing (creating) the 24 climate maps to represent the dynamicity of climate factors in four seasons as experienced by the marshes area in Southern Iraq.

2. Study Area

The area of study is located at the Mesopotamian Marshes (Figure 1). It was once the largest wetlands in the Middle East and Western Eurasia. The Marshes formed from annual flood pulses of the Tigris and Euphrates Rivers. In the 1970s, the Marshes covered between 15000 and 20000 km^2 of water surface and vegetation (UNEP 2001). The long term resiliency of the southern Mesopotamian landscape provided abundant resources to humans since the time of the Ubaid–Sumerian Culture (4000 – 6000 BCE) (Pournelle and Algaze 2010, Hritz et al.2012).The indigenous people of the Marshes, i.e. the Marshes Arabs have practiced sustainable traditional resource management for thousands of years, developing an iconic way of life that ties them intimately to their wetland landscape (Salim 1962). Traditionally, women took an active role both inside and outside the home, gathering reeds producing handicrafts, helping care for water buffalo, working in the fields, and selling in the market (Algaze, 2001). During the 1980s, the Marshes ecosystem and its inhabitants were subjected to large scale destruction during the Iraq–Iran War. Destruction continued during the 1990s in "one of the world's worst environmental disasters" when the previous Iraq government systematically drained over 90% of the Marshes in a military action designed to crush its inhabitants (UNEP 2001).With the fall of the previous regime in 2003, the Iraqi people began to re-flood the Marshes, and some Marsh Arab tribes returned to reclaim their lives. Good water years and Iraqi restoration efforts in 2004–2008 resulted in successful rehydration of approximately 58% of the Marshes (UNEP 2005, Al – Handel and Hu, 2015).Uncontrolled releases of water partially restored some areas, whereas restoration failed in other areas because of high soil and water salinity (Richardson et al. 2005, Richardson and Hussain 2006).The re-flooding and rehydration efforts since the fall of the previous regime have thus had highly variable results. Unfortunately, recent drought years, upstream water diversions on the Tigris and Euphrates Rivers, and climate change have desiccated large areas of the Marshes, resulting in Marsh inhabitants being displaced and without a livelihood.

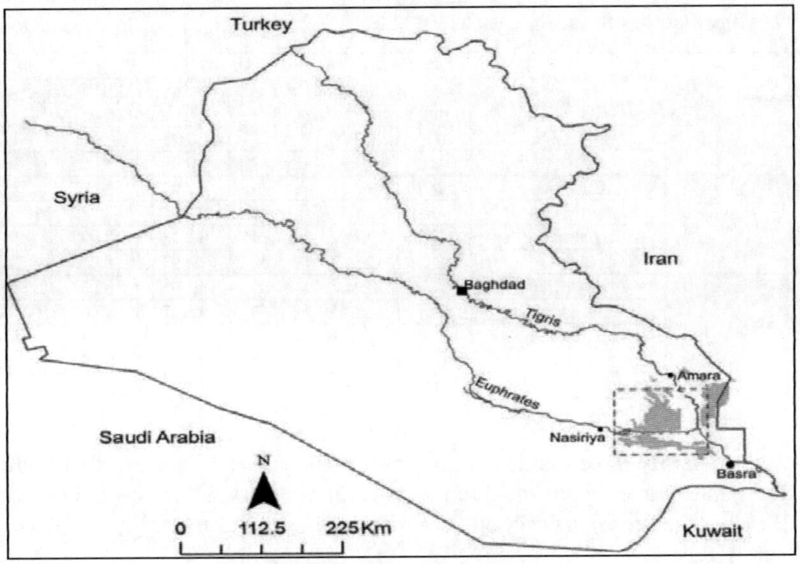

Figure 1 The Study Area

3. The Methodology

The combination of GIS and spatial interpolation technique is appropriate for the climate analysis. The proposed spatial model for multi season analysis was accomplished in GIS-IDRISI 17.0 Selva Version software. Other software utilities used in the analysis are ArcGIS 9 (ArcCatalog Version 9.3 and ArcMap Version 9.3), Adobe Photoshop 7.0 and Adobe Acrobat 9 Pro. Data acquisition of the four climate factors (temperature, rainfall, wind speed, and humidity) for four seasons in 2016 was obtained from the meteorological stations in the study area, that is (1 station) in Missan Governorate, (4 stations) in Thi-Qar Governorate and (4 stations) in Basra Governorate as shown in (Table 1). The map of the study area was prepared in hard and soft copy .

The hard copy map was initially resampled to register the correct coordinates. Subsequently, the coordinate locations of nine meteorological stations in the study area were determined and were linked to the database consisting the values of the climate data (temperature, rainfall, wind speed and humidity). Initially, the months for each season in the year (per year) was determined as follows:-

i. Winter Season - December, January and February.
ii. Spring Season - March, April and May.
iii. Summer Season - June, July, August and September.
iv. Autumn Season- October and November.

Table 1 Weather Factors for January, 2016

January - 2016									
ID	Governorates	Stations	R mm	TM C°	Tm C°	T C°	HM %	Hm %	WS m/s
1	Missan	Ali Al Gharbi	8.8	17.20	4.31	10.21	93.07	42.64	2.02
2		Kahla	1.5	17.38	5.17	10.98	87.05	38.91	1.22
3	Thi-Qar	Chabaish	2.7	19.64	7.43	12.94	86.67	35.18	1.55
4		Souk El Shuyouk	1.2	18.42	6.01	11.65	82.32	32.09	1.90
5		Shatrah	3.2	17.81	4.94	11.38	84.77	34.62	1.34
6		Fajr	6.2	17.63	5.00	10.93	83.72	36.20	1.74
7	Basra	Abu Al Khasib	7.3	19.03	7.82	13.01	85.34	35.99	2.49
8		Burjisiyah	7.7	17.65	6.43	11.61	85.67	35.35	2.40
9		Faw	6.4	18.21	7.84	12.80	82.69	37.28	2.98
10		Qurna	6.5	17.98	6.01	11.80	83.91	36.52	1.64

Subsequently, the average values of the four climate factors for each season were calculated. Since the temperature and the humidity have maximum and minimum values, therefore there will be an average of the minimum values and an average of the maximum values considered for each temperature and humidity in each season. Hence, each season has six climate maps representing four climate factors. For example, for winter season there will be:

1. Winter Average Max. Temperature Map.
2. Winter Average Min. Temperature Map.
3. Winter Average Rainfall Map.
4. Winter Average Wind Speed Map.
5. Winter Average Max. Humidity Map.
6. Winter Average Min. Humidity Map.

Similarly for other seasons like spring, summer or autumn will has the same number of maps. Therefore, there will be 24 climate maps of four climate factors for four respective seasons. These maps were derived in the GIS-IDRISI software and their values are then linked the nine meteorological stations.

Finally, converting the stationary 24 climate data into continuous climate maps requires the spatial interpolation process. The spatial interpolation estimates unknown values of climate data within the study area based on the data available at the respective meteorological stations. In this study, the Inverse Distance Weighting (IDW) algorithm available in GIS-IDRISI software was used to estimate cell values by averaging the values of sample data points in the neighbourhood of each processing cell. The weight of each sample is inversely proportional to the distance which means the further away from the point from the unsampled point, the less the weight in helping to know the values.

4. Results and Discussions

The output of this research has created total of 24 continuous climate maps of the study area showing the four seasons (winter, spring, summer and autumn) in the year 2016. It represents the four climate factors (temperature, rainfall, wind speed and humidity) interpolated from nine established meteorological stations. The Figure below described the respective maps.

In the Figure, the legend on the hand right side describes the range of values of each climate factor. The lowest value is on the top while the highest value is in the bottom. From these climate maps, spatial query or search using the cursor inquiry mode for any climate values within the study area is possible. The derived climate maps were intended for the application of an ongoing ecotourism model in improving and habilitating the marshes area in Southern Iraq. The intended study considers the climate condition in Iraq as dynamic variables that comprises of four seasons i.e. winter, spring, summer, and autumn that influences the climate factors such as rainfall, temperature, humidity and wind speed in each season. Lastly, the study will lead to the development of multiple seasons criteria site suitability model for ecotourism in the marshes area, Southern Iraq.

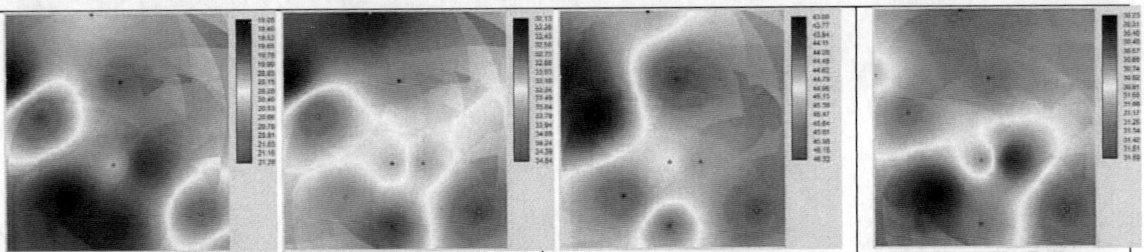

Winter Average Max Temperature Map	Spring Average Max Temperature Map	Summer Average Max Temperature Map	Autumn Average Max Temperature Map

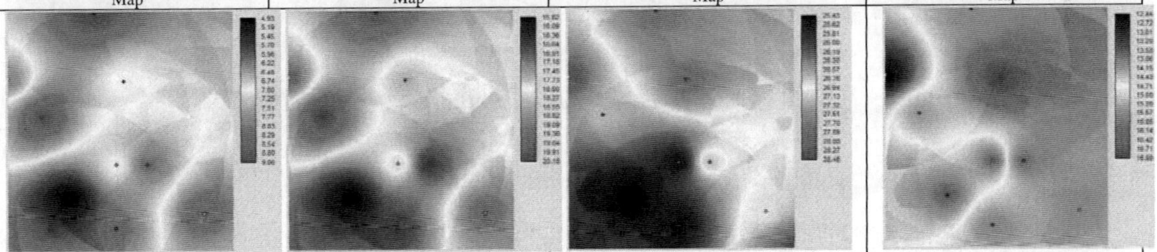

Winter Average Min Temperature Map	Spring Average Min Temperature Map	Summer Average Min Temperature Map	Autumn Average Min Temperature Map

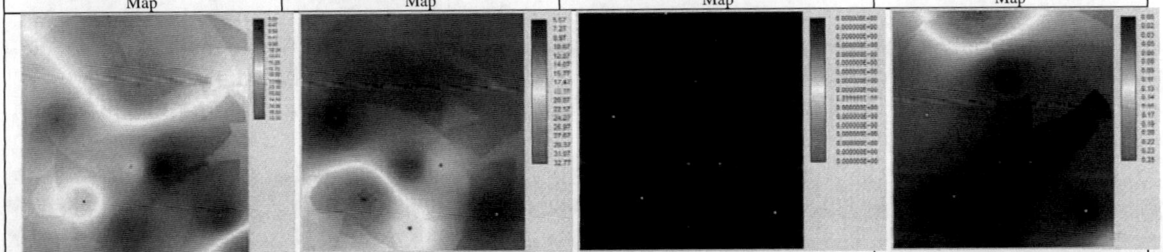

Winter Average Rainfall Map	Spring Average Rainfall Map	Summer Average Rainfall Map	Autumn Average Rainfall Map

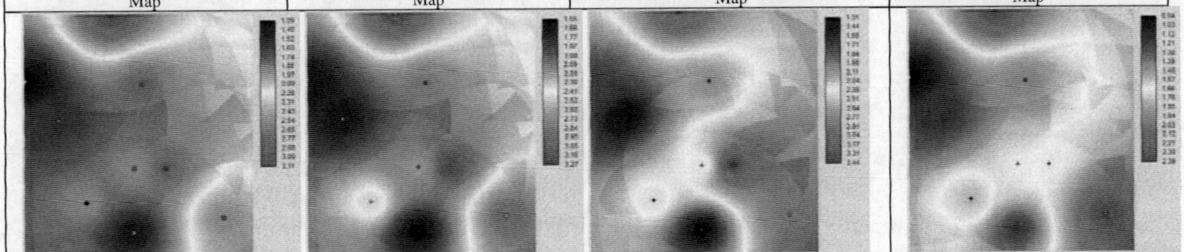

Winter Average Wind Speed Map	Spring Average Wind Speed Map	Summer Average Wind Speed Map	Autumn Average Wind Speed Map

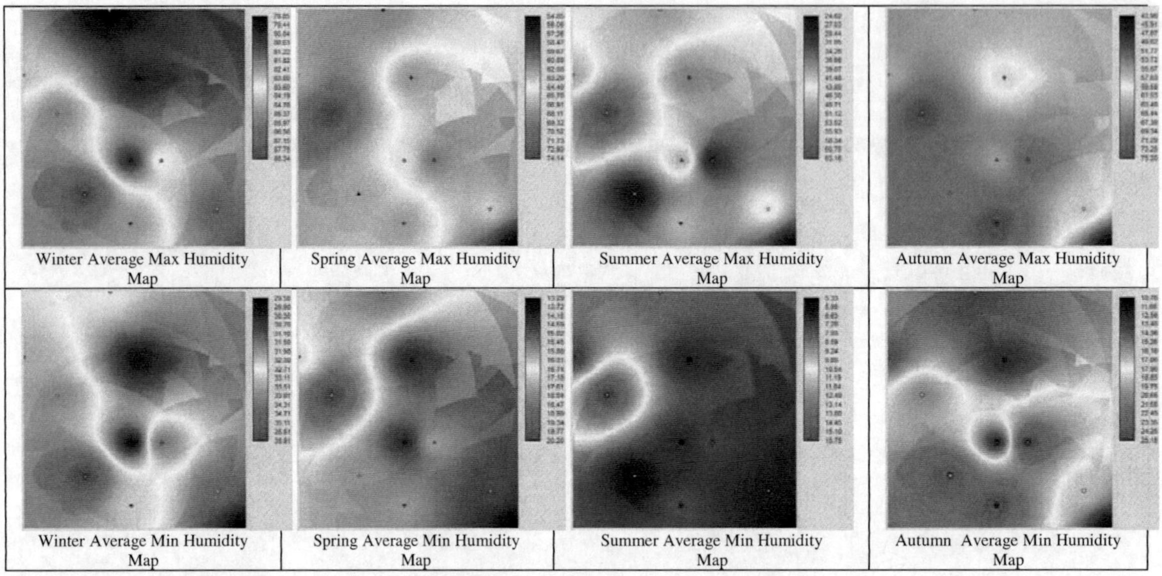

Figure 2 The 24 Climate Maps with Meteorological Stations on Them

5. Conclusions

- The climate maps of a particular area can be used for multi - purpose study either for development or for preparing a project such as tourism, ecotourism, the establishment of a factory, cultivation of certain crops…etc.
- The created dynamic climate maps for each required climate factor are in high resolution (36m x 32m) pixel by the Interpolation Method using GIS – IDRISI software .
- The GIS spatial analytical technique applied in this study can reduce time and cost by 20 - 30 % compared to traditional methods especially in the improvement and habilitation program where multi - purpose suitability model capability is commonly available in many GIS software.
- In addition , this method of obtaining climate factors values is the ideal and one of few methods in difficult areas where there are mountains , forests …etc. or in dangerous areas which have such as predators , remnants of war …etc.

Acknowledgement

I thank God for the strong will and determination in doing my PhD research. My acknowledgement to the School of Civil Engineering, USM and my supervisor for the research opportunity. I also thank the Director of the Climate Division at the Ministry of Agriculture in Baghdad for the provision of the climate data.

IOP Conf. Series: Materials Science and Engineering **401** (2018) 012014 doi:10.1088/1757-899X/401/1/012014

References

[1] Algaze, G.2001.Initial social complexity in Southeastern Asia: The Mesopotamian Advantage.Current Anthropology 42199 – 233.

[2] Al – Handel, A., and C.Hu.2015.MODIS observations of humaninduced changes in the Mesopotamian Marshes in Iraq.Wetlands 35: 31 – 40.

[3] Hritz, C., J. Pournelle, and J.Smith.2012.Revisiting the Sealands: Report of preliminary ground reconnaissance in the Hammar District, Dhi Qar and Basra Governorates, Iraq .Iraq 74: 37 – 49.

[4] NASA (National Aeronautics and Space Administration).2005.What's the Difference between Weather and Climate? NASA Television, NASA, Washington, USA .https: //www.nasa.gov / multimedia / nasatv /.

[5] Nicholson, E., and P. Clark.2002.The Iraqi Marshlands: a human and environmental Study .The Amar Appeal International Charitable Foundation, London, UK.

[6] Pournelle, J., and G. Algaze . 2010 .Travels in Edin: deltaic resilience and early Urbanism in Greater Mesopotamia. Pages 7 – 34 in

[7] A.McMahon and H. Crawford, editors. Preludes to Urbanism: the late Chalcolithic of Mesopotamia .University of Cambridge, Cambridge, UK.

[8] Richardson , C.J., and N. Hussain .2006.Restoring the Garden of Eden : an ecological assessment of the marshes of Iraq . BioScience 56 : 477 - 489.

[9] Richardson , C.J. , P.Reiss , N.Hussain , A.Alwash ,and D.Pool .2005.The restoration potential of the Mesopotamian marshes of Iraq .Science 307 : 1307 – 1311.

[10] Salim , S. 1962.Marsh dwellers of the Euphrates Delta .London School of Economics Monographs on Social Anthropology , Athlone , London , UK.

[11] UNEP (United Nations Environment Programme).2001.The Mesopotamian marshlands : demise of an ecosystem , early warning and ssessment. Early warning and assessment Division of Early Warning and assessment , United Nations Environment Programme , Nairobi ,Kenya.

[12] UNEP (United Nations Environment Programme).2005.Project to help manage and restore the Iraq Marshlands.UNEP Technical Report 47.UNEP, Nairobi, Kenya. http : // marshlands.unep.or.jp / .

IOP Conf. Series: Materials Science and Engineering **401** (2018) 012015 doi:10.1088/1757-899X/401/1/012015

The Beneficial of BIM Adoption in the Interior Design Services for Design Process

A B Abd Hamid[1] and M R Embi[2]

[1] Postgraduate Studies, Department of Architecture, Faculty of Built Environment and Surveying, Universiti Teknologi Malaysia, Johor Malaysia,
[2]Department of Architecture, Faculty of Built Environment and Surveying, Universiti Teknologi Malaysia, Johor Malaysia,

E-mail: abah7591@salam.uitm.edu.my

Abstract. The design process is a preliminary to design the space and development of the process to build a building by the further systematic layout and appropriately. Through the design phases, designers are capable to prepare the drawings and organize the documents using by Building Information Modelling (BIM). This technology and innovation be able to enhance the performance of interior design services by improvement of skill and knowledge. These researches are to classify the element of services in the interior design and to evaluate the effectiveness of BIM usage in design process for the interior design industry. BIM is imperative for industry players to be the best construction by technologies implementation. Through this study, the number of sixty-three interior design firms was chosen and limit to surrounding in Klang Valley registered with Lembaga Arkitek Malaysia (LAM). The survey was conducted through the questionnaire survey and the data collection will be analysed to grant significant information in BIM adoption for interior design services.

1. Introduction

Today, a new technology has continuously changed the approach of people conduct business and their lives which greatly impacted their daily life [1]. As an interior designer, there is no different to others accepted and incorporated with new innovation technology too has impacted the interior designers business, cooperation and communication with clients, and how to adopt new ideas and communicate [2]. As an interior design, the decision making is very important which is known as a design process. There is has several steps included the design idea, improvement of design, contract documents and contract administration.

Although designers have a design process, they are still to think the innovative design solution while fulfilling client's need. This process is defined as design thinking, solution issues-focused, problem-solving methods even as affording innovative approaches to explore and analyze design solutions [3, 4, 5, 6, &7]. There is a need too to beyond approach of the interior design industry and glance toward human interactive or environment interactive conceptual and theoretical element. The design solution is generally significant to accommodate in the design phases to be efficiently processed for the construction and documentation by implement BIM for interior design firms. Interior design can increase the knowledge and develop a skill in the services for the interior design industry with BIM adoption.

Content from this work may be used under the terms of the Creative Commons Attribution 3.0 licence. Any further distribution of this work must maintain attribution to the author(s) and the title of the work, journal citation and DOI.
Published under licence by IOP Publishing Ltd

2. Literature Review

2.1. Interior Design Services

An interior designer has enclosed on training, space planning, construction, detailing, and finishes materials selection in addition to considerations of furnishings. In the construction and design, it is describes client and user requirements, develops on conceptual approach, and creates documents solutions. In the interior design, project management is a part of services provided to fulfill projects demands, knowledge and skilful [8], even though as a interior designers should be acknowledgeable in term of design criteria and must not be left behind in the technical skills and requirements of projects with interpersonal communication and management strategies flow together with the technology updated [9].

However, interior designer's job scope not limited only design; the capacity of interior designer's work similar to architecture and of course is yet it to limited process to the enclosed space either design tasking and processes of drawing. Not only talking about design, interior design also involved in the construction and completion phases of projects involved. The process can only be successful by conducting through proper planning, executing the design, tremendous monitoring and fair controlling until the project hand over.

2.2. A Theory and Practice of BIM

BIM is a process for design practice and standards of building design, delivery and operation. The intention of BIM does not only spotlight on traditional method like CAD but is more an inventive and indicator of innovation that is enormously smart to tolerate for a proper management in design [10]. Implement BIM can enhance the organization of human capabilities and organization by declining interruptions and imprecision between a clients.

BIM endowment the opportunities to cooperated with each other and to employ the data in the certain areas, values, and commercial aspect in construction industry [11, 12]. Furthermore, technology have a capability to contribution the competitive advantage by improve chances and decrease the costs [13] that is showing as a function of Technology, Process and People [14]. Interior design process is merging with some design decisions, occupied and improved the driven by a miscellany of factor; space consumption, area conditions, visuals, material rate, finishes and material – to ensure for decision-making development. BIM adoption using via technology, the documentation every phases will increase a efficient process to enable the speedy movement of the project.

2.3. Interior Design and BIM

Interior design is a professional practice with imaginative and solution finder to applied the possession of field to attain a superior environment. Solving the problems can increase the quality of life and occupied culturing with aesthetically striking. Based [15] it is mentioned that illustrate of the interior designers are needed a good thoughtful on construction aspects to certify effectiveness using BIM. With the innovation and new technology, interior designers are obliged to comprehend the core and implementation of building details in design effectively in BIM.

Interior designer’s portrays a building's as the inner space with the concrete scale to manipulate the design components for instance wall, ceiling and floor. Interior designer's has to distress about the human ergonomic to certified the working and purposeful space. According to [16], interior design is expertise body to be a creative thinking and technical skills to solve design problems.

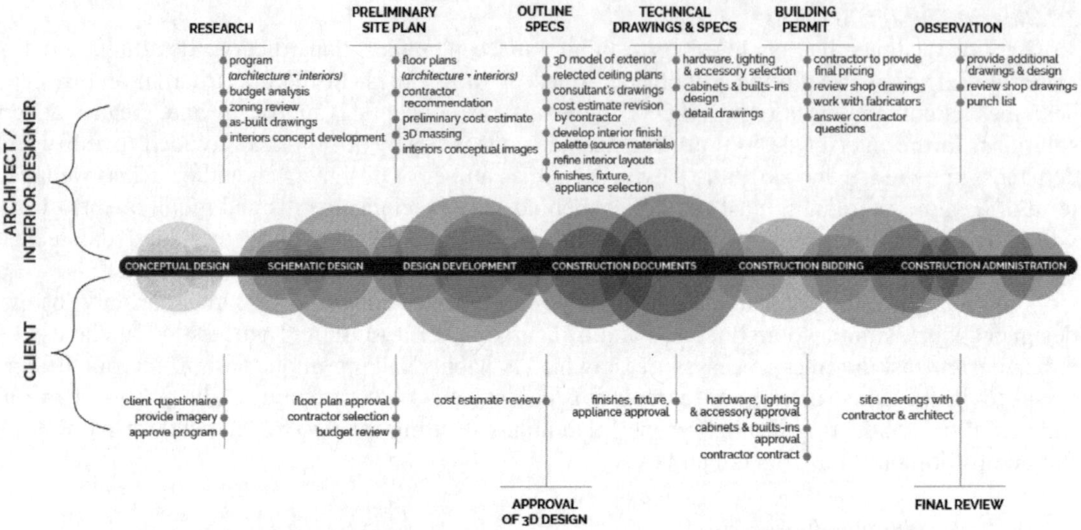

Figure 1. The design process work flow. (http://hmhai.com/design-phases

Table 1. Knowledge of BIM for Interior Design

BIM for Interior Design		
Interior Advantages	1. Speed and ease of interior design creation and its ability to visualize the design 2. Ability to capture and manage the design via multiple options – from space layout to material selections. 3. The richness and reliability of data - early fundamental task.	
Interior Design Options	1. To keep multiple design alternative until enough information decision making. 2. Enabling designers to develop and study multiple design alternatives simultaneously within a single model.	
Interior Design Information	a. Schematic design	- Able to create a master schedule for interior spaces. - A working database view of a project. - Able to input table-based schedules for many interior design components. - Produce accurate preliminary estimates.
	b. Calculating Quantities	- Storage, manage and coordinates design - Calculate area and quantify material and reliably – reducing overall project cost - Accurate area measurement – carpet, wall coverings, ceiling tile, or finishes. - Ability to track occupancy, space allocation and assets.
	c. Interior Documentation	- All the construction documents relating to the interior design are direct presentation.

3. Research Aim and Methodology

This research aims to identify and evaluate the benefits of BIM for interior design industry in Malaysia in the design process. This research is to find the data about the benefits and barriers to adopt BIM in the interior design industry. This research focusing on interior design firms registered with Lembaga Arkitek Malaysia(LAM) and the total of number sixty-three interior design firms was chosen in this study as a case study.

IOP Conf. Series: Materials Science and Engineering **401** (2018) 012015 doi:10.1088/1757-899X/401/1/012015

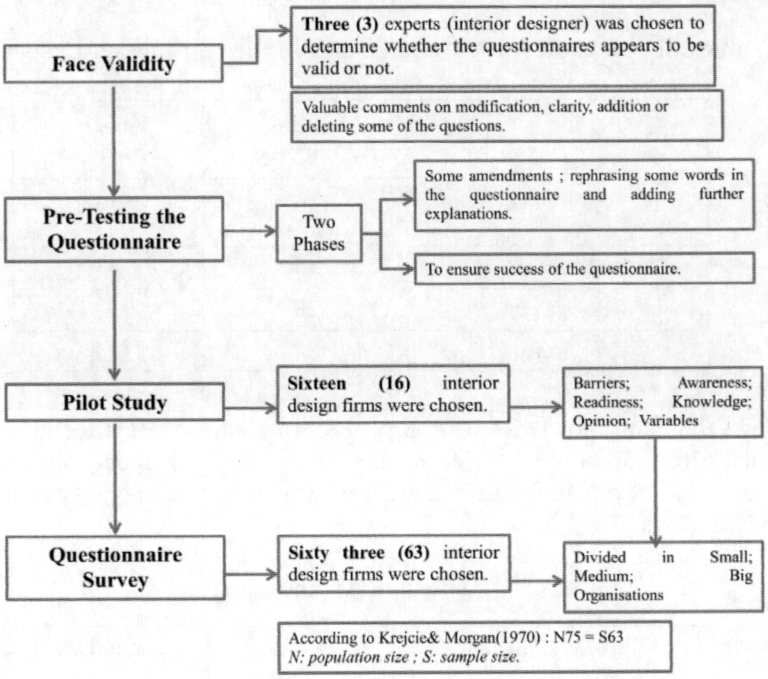

Figure 2. Research Method Process

4. Result Analysis

According to the result, this study focusing on the beneficial BIM adoption to interior design firms in the design process and identifying the awareness, readiness, problems and barriers of BIM appropriate usage in the interior design firms.

According Table 2 shows only 11 respondents firm was established in year 1990 and below while 10 respondents firm was established between 1991 and 2000. Meanwhile, 19 respondents firm was established between 2001 and 2010. However, only 23 respondents firm was established in year 2011 and above.

Table 2. Interior Design Firms Established

Year	Size of Interior Design Organization		
	Small firms (*f*)	Medium firms (*f*)	Big firms (*f*)
1990 and below	3	5	3
1991 - 2000	5	3	2
2001 - 2010	8	5	6
2011 and above	11	8	4
Total	**27**	**21**	**15**

Table 3 shows the awareness and readiness level of interior design using BIM. The level of awareness and readiness for small firms are higher than the big firms with different percentage 19% compare medium firms was 33%. It is shown; small firms are ready to used BIM comparing two others size of organisation in the interior design firms.

Table 3. Level of Awareness and Readiness BIM for Interior Design

Size of Interior Design Organization (*f*)	Level of Awareness	Level of Readiness
	%	%
Small firms (*f*)	43	43
Medium firms (*f*)	33	33
Big firms (*f*)	24	24

Table 4 shows the BIM knowledge instead of the experience of BIM for interior design firms. From the table, small firms are very advanced for knowledge with 9 respondents comparing medium and big firms. Small firms also lead for knowledge zero which is 18 respondents answering compare medium forms and big firms.

Table 4. BIM Knowledge

BIM Knowledge	Small firms (*f*)	Medium firms (*f*)	Big firms (*f*)
Yes	9	5	3
No	18	16	12

Figure 3 shows the services for interior design in Malaysia and the most popular are software applications with 15 respondents agreed with BIM in the software application. The next is with 13 respondents are interior design documentation and selection of colour most agreed to have in BIM adoption.

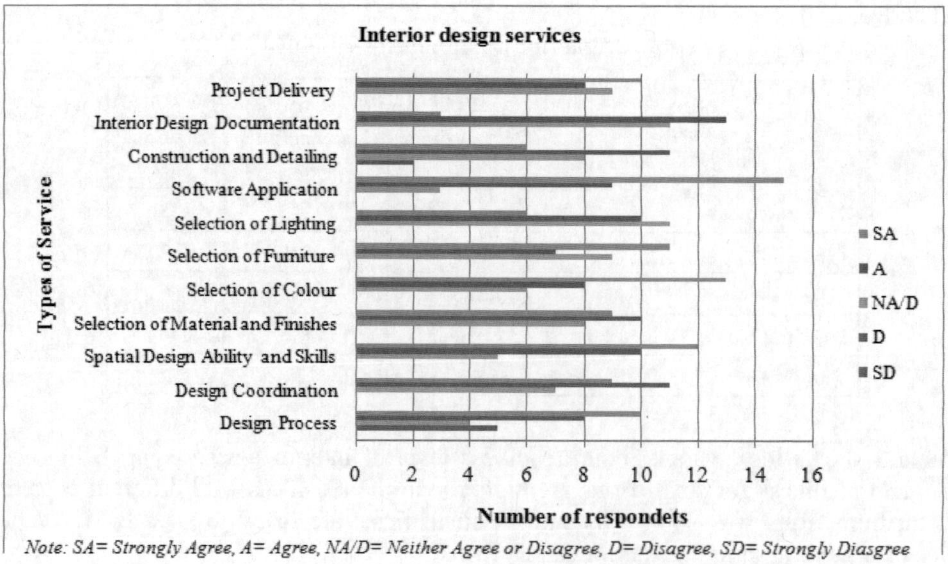

Figure 3. Interior Design Services

5. Discussion and Conclusion

Implementation BIM for interior design industry still need to improve the adoption and awareness. BIM technology can help interior designers to find the solution on design and design process while several factors need consideration in BIM implement. Knowledge and financial are the main barriers to ensure BIM implement in the interior design industry. Beneficial of BIM should be considered to ensure interior design not to behind from others construction industry and must have to move ahead in term of technology and innovation in the construction industry. Ensuring BIM will be implemented in interior design industry, government and organisation related to the interior design body must have manage and organise to develop and enhance application BIM in the interior design industry.

Acknowledgments

Authors would like to acknowledge to interior design firms and to who assist along the research progress for contributing part of this research.

References

[1] Lawson B 2006 *How Designers Think: The Design Process Demystified*. Burlington MA: Architectural Press.

[2] Kalay Y 2006 *The Impact of Information Technology on Design Methods, Products and Practices*. Design Studies *27*, 357-380.

[3] Adams R Daly S Mann L and Dall'Alba G 2011 *Being a professional: Three lenses into design thinking, acting and being* Design Studies **32**, 588-607.

[4] Cross N 2006 *Designedly Ways of Knowing* London: Springer-Verlagl

[5] Lande M and Leifer L 2010 *Incubating engineers, hatching design thinkers: Mechanical engineering students learning design through ambidextrous ways of thinking*. In Proceedings of the Annual American Society for Engineering Education Conference, Louisville, KY.

[6] Lawson B and Dorst K 2009 *Design Expertise*. Oxford: Architectural Press Lauer, D.

[7] Pentak, S 2008 *Design Basics (7th ed.)* Boston, MA: Thomson Wadsworth.

[8] Piotrowski C M 2008 *Professional Practice for Interior Designer. (4th ed)* New Jersey: John Wiley & Sons, Inc.

[9] Nur Maizura A N 2012 *Interior Design Project Manager Competency*. Unpublished PhD draft thesis. *Universiti Teknologi MARA (UiTM), Shah Alam*, Malaysia.

[10] Karen M K 2014 *Introduction: On the Theory and Practice of BIM, Building Information Modeling: BIM in Current and Future Practice*, John Wiley and Sons Inc Hoboken New Jersey

[11] Tse T K Wong A K and Wong F K 2005 *The Utilisation of Building Information Models In nD Modelling: A Study of Data Interfacing and Adoption Barriers, Electronic Journal of Information Technology in Construction*, **vol. 10**, pp. 85-110.

[12] Gao J and Martin F 2008 *Framework & Case Studies Comparing Implementations & Impacts of 3D/4D Modeling Across Projects*. Technical Report Number **172** Centre for Integrated Facility Engineering, University of Salford, UK.

[13] Mitsutaka H 1989 *Evaluation of Technology in Construction*. Technical Report Number **16** Centre for Integrated Facility Engineering, University of Salford, UK.

[14] Koskela L 1992 *Application for the New Production Philosophy* Technical Report Number **72**, Centre for Integrated Facility Engineering, University of Salford, UK.

[15] Crumpton A Miller B 2008 *Building Information Modeling: State of the A&D Industry and BIM into Design Education*. In Proceedings of the Interior Design Educator's Council Annual Conference 1175–1188.

[16] National Council for Interior Design Qualification 2000 *NCIDQ Examination Study Guide*. Washington, DC: NCIDQ, pp. 22–23

An Architectural Characteristic and Local Conservation Study of a Malay Traditional House at Kampung Terusan, Kuala Pilah.

Mior Zainal Abidin, Norhati Ibrahim, Mastor Surat.

National University of Malaysia, 43600 Bangi, Malaysia.

miorzainalabidin@gmail.com

Abstract. The house of Al-marhum Raja Kamarudin Raja Yaa'cob, estimated to be 160 years old, can still be seen standing and well-maintained even though it had been inherited for four generations. All its structural members which are pillars, beams, floor joists, wall plates, and rafters made of hard wood timber called 'Cengal Batu' by the locals, however the non-structural members which is wall panels, window panels, and floor boards are seen to be weakened by termites. Some part of the non-load bearing walls has been replaced to new and different type of timber thus making it less durable that of the structural timber. For the residence of Kampung Terusan, their individual land plot is divided by agricultural and residential plot. These relation of land and houses is the main factor why Kampung Terusan is densely populated in the days before public water and electrical supply had reach the village. Here, the sustainability of the area was analysed based on interviews and observations. Traditional houses in Negeri Sembilan are unique particularly on the roof shape and structure. It uses less rafters as the roof coverings are of light materials, are among factors studied and explained using redrawn drawings, interviews from architects and observations.

Keywords: Malay Traditional House, Kuala Pilah,Sustainability, Minangkabau house, Conservation, Timber.

1. Introduction

At current time, the house of Raja Kamarudin is among the few houses in Kampung Terusan that still has its pillars, beams, and window panels are still in good condition and structurally strong of taking current load, but through observations, the timber wall of the rear left is almost 50 % were taken by termites. Even at this condition, the house is still standing straight without any tilted pillars which can be seen on some other houses of its same nature in Kampung Terusan. The location of this house now is not on its original place but had been moved from where it is constructed 160 years ago. This shift which takes place in 1993, marks the move of conservation by the owner hence replacing deteriorated timber member. The fact that this house remains at its original place for 135 years was analysed to find out why it has sustained that long and is there any possibility that it might stand for a few generation to come. Main clue to this is that Kampung Terusan is a village of significant historical background which

Content from this work may be used under the terms of the Creative Commons Attribution 3.0 licence. Any further distribution of this work must maintain attribution to the author(s) and the title of the work, journal citation and DOI.
Published under licence by IOP Publishing Ltd

IOP Conf. Series: Materials Science and Engineering **401** (2018) 012016 doi:10.1088/1757-899X/401/1/012016

goes back to the 17th century when the Minangkabau tribe that is said to flea disturbance at their home town in Sumatra, and opened land for paddy field after marital engagement with the locals (1).

Ancient findings of megalith structure (grave stone) that can be seen at two Islamic graveyards are among other facts to prove that the village setting has sustained for a millennium. These settlers would also influence architectural design by bringing their culture and religion to Kampung Terusan. Possibilities of what micro climate might have inspirited architectural design or not, apart from material availability and sustainability, also to be studied in finding house of Raja Kamarudin's architectural characteristic.

2. Materials and Methods

Research methodology for this thesis are based on observations, interviews, and literature reviews. Observations were done first to identify the evidence of structural and non-structural strength of selected house. The search started with redrawing plan and elevation of the house by measuring at current state. Data collection of measurement than are extended by examining structural timber condition and then recorded visually. Parts of the house that has been replaced are also identified noting year of change and current technology. Secondly, the interviews are used to find year of built, origin of owner for architectural influence, and conservation method and process. And third, literature reviews is to search for clues on the evidence of early settlement in Kampung Terusan.

Figure 1: Views of Megalithic Islamic Grave structure at Kampung Terusan.

Figure 1 shows a few megalith graveyard stones that are also located within the compound of current Islamic cemetery. Megalith structure can be traced back to about 2000 years ago, but Islam is said to had arrive in the Malay Peninsula during the 10th century (2), thus marking the date of those megalithic grave stone in Kampung Terusan. Being of 1000 years old, there are the only structure that still can be

IOP Conf. Series: Materials Science and Engineering **401** (2018) 012016 doi:10.1088/1757-899X/401/1/012016

seen unlike houses, where the materials are mostly timber which life time is about 100-180 years old depending on the maintenance.

The interview with the current owner of Raja Kamarudin's house was mainly to investigate the architectural history of it. "Rumah ini dulu duduk di baroh. Masa tu tak ada air paip, pakai air perigi untuk mandi dengan air minum." (3). The position of this house originally was at the 'baroh' (land beside paddy field and Sungai Muar), which there was no water and electrical supply yet, and we have to use water from the well for drinking. Kamarudin explains the original location of his house and pointed that it there so that its nearer to work the paddy fields. He added, "...ada kemudahan air paip dengan letrik (1976), mula-mula masuk kat tanah baroh, dalam 10 tahun kemudian (1986), arwah Abah pun buat rumah batu kat tanah darat. Agak-agak dalam 20 tahun lepas pindah rumah darat (1993), Abah mintak tolong tukang bukak dinding, atap dengan lantai, lepas tu angkat ramai-ramai ke darat dengan bantuan orang kampung, sebab masa tu dah tak ada orang tinggal kat baroh dan Abah takut rumah tu kena makan anai-anai" (3). ...water and electricity were supplied to our house at 'baroh'(1976), 10 years after that our late dad built a new house (masonry) at 'darat' and we moved in. In 1993, we brought the wooden house to the front by disassembling it walls, floor and roof, and carry it to the new location (400 meters away). Here, Kamarudin note the move of the house was because so it is nearer to be maintained. "...tiang rumah lama ni orang dulu cakap jenis 'Cengal Batu'. Kayu ni jenis yang tak makan paku, anai-anai pun tak mahu makan, tu sebab pakai pasak"(3). The pillars of this house were made timber called 'cengal batu'. This timber is so strong that nails could not penetrate and termites wont' eat, thus the builders is using 'pasak' (piece of small round timber used to lock joints. "...sekarang ni adik saya yang bongsu (Raja Norziaton) duduk situ, rumah kayu tu dia la pulak yang menjaganya"(3). ...and now my younger sister Raja Norziaton Raja Kamarudin is staying at this house and it is up to her now how she would preserve it. Kamarudin marks another conservation point of this traditional Malay house.

"...saya tahu rumah tu, dulu di baroh, masa nak pindah ke darat, saya ada sekali menolong. Orang kampung datang dalam 20 orang. Mula-mula kami buka dinding, lantai dan atap. Kemudian kami angkat ramai-ramai guna kayu, dalam 3 orang untuk satu tiang dan kami letak kat darat. Lepas tu tukang pasang balik dinding, lantai dan atap ikut keadaan asal"(4). ...I know that house, it was at 'baroh' originally and I was one of about 20 villagers that helped moving the house. First, we dissemble the walls, floors and roof, then we carry with 3 people at a pillar to its current location. Once there, the carpenter would reinstall the parts of the house. Sufian is a neighbour of Raja Kamarudin.

Facts on architecture influence were taken from Samad's writing on houses in Negeri Sembilan as, "Di Sumatera, terdapat rumah -rumah tradisional yang sengaja dibuat lentik dinamakan rumah lontik. Bumbung yang lentik merupakan rekaan yang berdasarkan konsep atau imej perahu. Atap perahu Melayu kelihatan seperti lentik yang disengajakan, iaitu bukan lentik alami yang disebabkan oleh graviti"(5). In Sumatra, there a trace of houses that uses a long curvy roof and were called 'lontik' (curve). In his opinion, the curvy roof was derived from the shape of a 'perahu' (small Malay wooden boat), and it is not as the force of gravity. "Bagi masyarakat Minangkabau, bumbung yang lentik merupakan simbolik kepada tanduk kerbau dan menjadi simbol kemenangan bagi masyarakat ini."(5). For the Minangkabau community, the curvy roof is a symbolic mimicry of a bull's horns which also portrays victory. Samad states another architectural influence of these Minangkabau roofs. He added on the interior space, "Semua rumah tradisional di Negeri Sembilan mempunyai susun atur ruang yang sama; berserambi panjang pada bahagian hadapan, ruang rumah ibu di tengah dan ruang rumah dapur di belakang.i- Ruang depan - dominan lelaki dan tetamu,ii- Ruang ibu rumah - ruang paling penting terhormat, iii- Ruang belakang - dominan perempuan dan tempat masak-memasak"(5). All Malay traditional houses in Negeri Sembilan has the same space zoning, long veranda at the front, 'rumah ibu' or living and bedroom in the middle, and kitchen at the back. i: front of the house – male and guest dominant space. ii: 'rumah ibu' – most important and private space. iii: rear space – dominant for women and cooking. Here, Samad is describing the space zoning of public and private space.

Figure 2: Views of Raja Kamarudin's house and its joint of pillar, beam and floor joist (lower right)

Figure 2 shows the elevation of the house. Its façade has a clear Minangkabau identity by just having the slender curve on the roof. No other curving element can be found on any other part of the house thus showing the strong dominance visually.

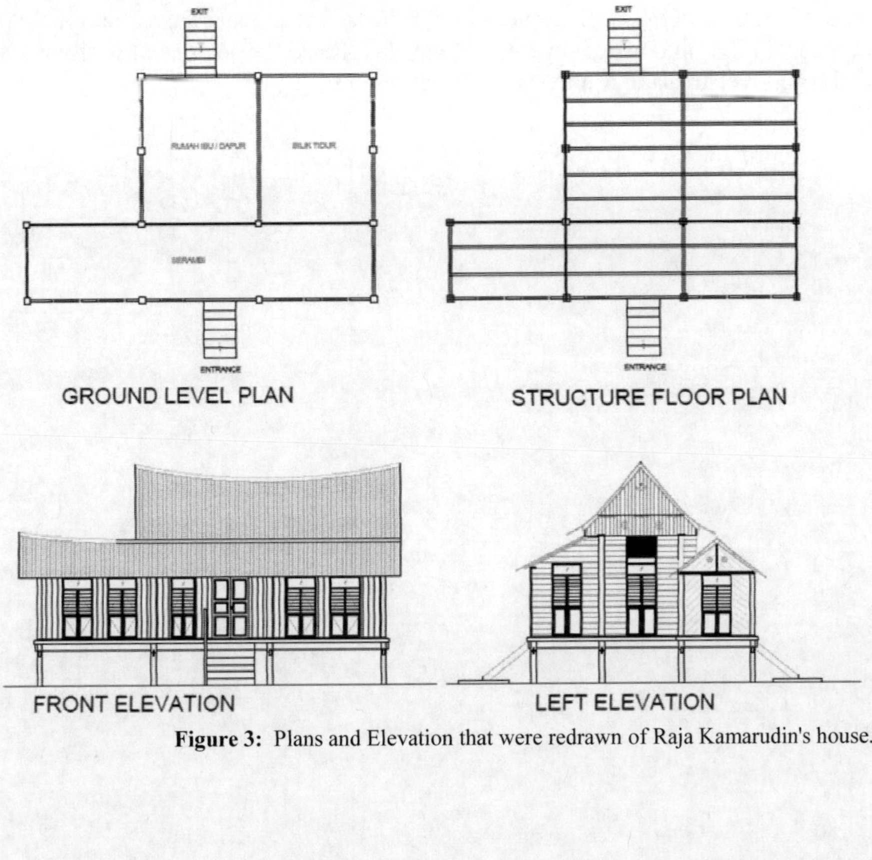

Figure 3: Plans and Elevation that were redrawn of Raja Kamarudin's house.

IOP Conf. Series: Materials Science and Engineering **401** (2018) 012016 doi:10.1088/1757-899X/401/1/012016

Figure 3 shows drawings that was used to study proportional and ergonomics of space and façade of the house. On the front elevation, roof curvy lines that is said not because of gravity pull, expresses motifs of the Minangkabau Community but is unique to Negeri Sembilan, as those in Sumatera has extreme curves and very pointy.

3. Results and Discussions

Microclimatic influence in architecture can be clearly seen through peninsula Malaysia on the different types of space planning, and roof design and angle. But for the Minangkabau community, their architectural influence has a mixture of cultural background imbedded architecturally. This can be proved in the curving roof (motif of boats and bull's horns), and in the space zoning where the Malays segregate private and non-private space to cater for guest of the house.

4. Conclusions

Based on the historical background of Kampung Terusan and Raja Kamarudin's timber house, it can be concluded that the value of conservation is always the priority of the owner even when new technology of preserving timber is far reach. The methodology of conservation in before chemical mix are used, remain unknown as the generation has passed. Nonetheless, the villagers thinking of that a house, will not deteriorate if it is occupied, might be the answer to timber houses that lasted for generations like the house of Raja Kamarudin.

5. References

[1] C.W.C. Parr WHM. Rembau, One of the Nine States. Its History, Constitution, and Customs. Journal of The Straits Branch of The Royal Asiatic Society. 1910;56(December 1910):2.

[2] Ahmad AS. Sulalatus Salatin: Sejarah Melayu. Kuala Lumpur: Dewan Bahasa dan Pustaka; 1979.

[3] Kamarudin RSR. Rumah kayu lama milik anda. In: Abidin MZ, editor. 2017.

[4] Sufian E. Perpindahan Rumah Raja Kamarudin dari Baroh ke Darat. In: Abidin MZ, editor. Kampung Terusan2017.

[5] Abd. Samad bin Kechot ZbH. Keunikan Seni bina Rumah tradisional Negeri Sembilan : Penilaian mengenai nilai-nilai artistiknya. In: Malaysia UK, editor. Warisan (Senibina dan Budaya). Universiti Kebangsaan Malaysiaunknown.

IOP Conf. Series: Materials Science and Engineering **401** (2018) 012017 doi:10.1088/1757-899X/401/1/012017

Efficiency of House-Walls Construction Using Building Ruins

Andi Prasetiyo Wibowo

Architecture Department, Faculty of Engineering, Universitas Atma Jaya Yogyakarta

e-mail : andiprasetiyo@staff.uajy.ac.id

Abstract. The stage of reconstruction of houses collapsed and damaged by the quake could amount to hundreds of thousands of homes and had to be rebuilt at almost the same time. This will result in a large need for building materials, thus affecting the price of building materials, as well as the demand of builders / construction workers. A major earthquake, such as the one that occurred in Yogyakarta, on May 27, 2006, has resulted in hundreds of thousands of homes severely damaged and collapsed, resulting in a large number of ruins of building materials in the form of masonry walls. The existence of waste material collapses, is a huge potential when reused as a building material through appropriate recycling technology, which can answer two problems at once that meet the needs of building materials and on the other hand can clean up the waste collapse of the building. The construction of walls from wall-crumbling powder (recycled material) is more efficient and effective than conventional wall-making systems when used for mass construction (more than 20 homes) at the same time.

1. Introduction

Green Technology is now one of the most talked-about issues around the world. Many parties try to develop it in various fields including in the construction world. The purpose of this concept is to answer the environmental problems that are increasingly leading to the destruction, which among others is the result of massive exploitation of natural resources on earth for construction, such as sand, stone, gravel, clay for brick making etc.

On the other hand, the use of very large amounts of construction in the world, for example : the construction of houses, buildings, bridges, dams will affect in a continual taking building material from nature that can lead to extinction of those material. This will disrupt the balance of the environment; the environment becomes damaged, which will ultimately harm human beings themselves.

Furthermore, we need an alternative way as a breakthrough step in preventing the occurrence of environmental damage is getting worse and bigger. Currently known as a sustainable development concept which offers a balance between the preservation of nature with the fulfillment of human needs that are growing in the future. According to World Commission on Environment and Development [1] in 1987, Sustainable is "meeting the needs of the present without harming future generations to meet their needs." There are three things that the goal of this concept is to achieve:

1. Minimize material and energy consumption.
2. Prevent negative effects on the carrying capacity of the environment and the environment itself.
3. Meet human needs.

Sustainable buildings are buildings that use methods and building materials that highly prioritize environmental quality, economic vitality and social benefits through the design of development,

Content from this work may be used under the terms of the Creative Commons Attribution 3.0 licence. Any further distribution of this work must maintain attribution to the author(s) and the title of the work, journal citation and DOI.
Published under licence by IOP Publishing Ltd

building operations, maintenance and deconstruction of the built environment. Standing buildings that use technology today are responsible for consuming 40% of the flow of energy and materials from the earth (Roodman & Lenssen[2]) so the effects of building these buildings are enormous for nature.

The recyclable material referred to in this paper is derived from the remains of unused building debris which is then processed in such a way that it can be reused for development. However, there are some things to consider before applying this concept, because sometimes the use of a technology if not properly studied, it will actually bring bad impact even worse than previous technology, among others, the increased energy needed to collect and process the material, or it turns out there is no environmentally friendly technology that can process the building materials, or even the fact that building materials derived from nature require energy and costs much smaller than the existing recycled materials. It is therefore necessary to select the correct and appropriate technology in the processing of recycled materials so that it can be reused for development.

One application of the use of the concept of recycled materials, it would be very appropriate when used in times of natural disasters that are currently often experienced, especially earthquakes, i.e. at the stage of reconstruction of houses collapsed and damaged by the earthquake, whose number can reach hundreds of thousands of homes, which had to be rebuilt at almost the same time, resulting in a high demand for the need for building materials, affecting the skyrocketing prices of construction materials, as well as the demand for builders / construction workers.

This can be seen, among others, on the occurrence of a major earthquake in Yogyakarta, on May 27, 2006 which has resulted in hundreds of thousands of homes suffered heavy damage and collapsed, resulting in waste of a large number of collapsed building materials in the form of a wall of masonry as shown in figure 1.

Figure 1. Waste of the wall ruins that are considered unusable. (Satyarno[3])

The local government of Yogyakarta dumped out the building ruins not on the appropriate land, some of which even dumped into places or productive land such as rice fields, due to limited costs. (Satyarno[3]; Satyarno[4]; Tjokrodimuljo[5]).

The existence of building ruins, is a huge potential when reused as a building material through appropriate recycling process, which can answer two problems at once that meet the needs of building materials and on the other hand can clean up the waste collapse of the building.

2. Research Methodology

The method of writing this paper is done by collecting secondary data through literature study and document study that is data collection based on the results of research related to the utilization and use of materials used building materials, especially waste building debris.

The data obtained will be analyzed qualitatively i.e. the analysis done by understanding and assembling data that has been collected and arranged systematically then drawn conclusion.

3. Analysis And Discussion

Waste disposal materials (wall ruins) can be processed by manual (human power) or using a crushing machine (Satyarno[3]). The crushing machine of the wall ruins as shown in Figure 2. In general, waste crushing machine works like stone crusher consisting of three parts, namely the frame of the cradle equipped with wheels for the machine to be easily transferred, the chamber/crushing room and a propulsion engine in the form of a diesel engine. Inside the chamber there are spinning blades to crush the waste of the wall ruins that poured into.

Figure 2. The ruins of the building walls were crushed using a machine. (Satyarno[3])

From the analysis of grain size analysis or gradation of fine aggregate showed that the powder of the wall ruins was in zone II (somewhat rough) with fine modulus of grain (MOG) of 2.61 where sand generally has MOG value between 1.5 to 3.8. The weight of the powder unit from the wall ruins is 1,360 gr / cm³ while the sand is 1,656 gr / cm³. From these results, the weight of powder units from the ruins of the wall is smaller than the sand but still includes the normal fine aggregate.Examination of specific gravity in the state of SSD (Saturated Surface Dry) obtained a value of 2,315 for powder from wall ruins and 2,669 for sand. Specific gravity in the absolute dry state of 2,082 for powder from wall ruins and 2594 for sand. From these results it can be seen that the powder density of the wall ruins is smaller than the sand but still includes the normal fine aggregate.

Utilization of the powder of the ruins of the wall as a building material based on the results of previous research (Satyarno[3]; Satyarno[4]), it is known that the powder ruins of the wall have higher absorption than natural sand, because according to the examination of chemical content in powder ruins wall has silica although the value is only 21.458%, so it is likely to have pozzolan properties. Thus a wall collapse which is then processed into a powder can be used instead of natural sand. According to the results of the research (Satyarno[3]; Satyarno[4]), powder wall debris when used as a substitute for sand and used as fine aggregate on the mortar mixture obtained the result that the mortar weight per cubic meter is smaller than that of sand mortar as fine aggregate.

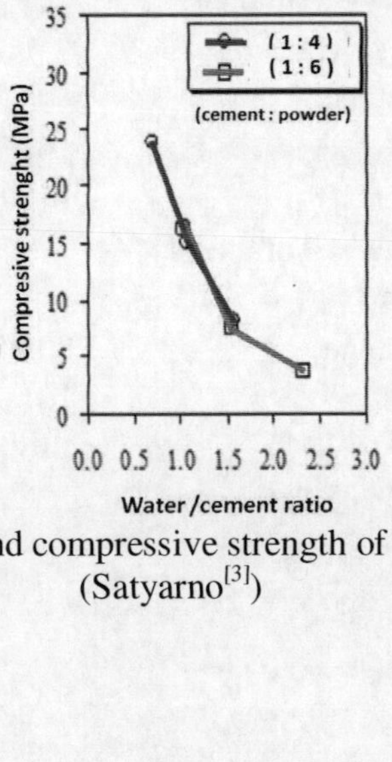

Figure 3. Portions of the mixture and compressive strength of mortar using powder wall debris (Satyarno[3])

From the test results (figures above), the compressive strength is obtained depending on the ratio of cement volume and powder depending on the portion of water used and it is recommended to use the ratio of cement volume: powder between 1: 4 to 1: 6 or can be used on average mixture of 1 cement: 1.25 water: 5 powder. The maximum gain strength can get 10 MPa.

3.1. Field Implementation (onsite)

The new wall-making system at the site uses reinforced concrete using a special mold instead of masonry. The material used to make the wall is a mortar with a mixture of water, cement, powder wall debris and inside the walls are fed a diameter of 6 mm diameter with a distance of about 40 cm.

Figure 4. The process of making the walls with cast in place method (Satyarno[3])

Price Comparison / Cost of Making Wall between using Special Method and conventional method, either based on SNI Analyst, and BOW Analyst. The advantages of financing the construction of wall work as described above, can be seen comparing the calculation as follows:

Table 1. Comparison of wall-making cost calculation

No	Method	Volume	Unit price (Rp)	cost (Rp)
1	Cast in place			
	a. 1 m2 Crushing Brick Waste	80.85	2,657.31	214,843.51
	b. 1 m2 wall casting	80.85	57,406.21	4,641,292.08
	c. 1 m2 mold installation	40.4	2,700.00	109,080.00
	Total cost		**4,965,215.59**	
2	Conventional, SNI Method			
	a. 1 m2 Brick Installation 1/2 stone	80.85	48,490.00	3,920,416.50
	b. 1 m2 Plastering	161.7	16,215.00	2,621,965.50
	c. 1 m2 Edging	116.8	3,458.00	403,894.40
	Total cost		**6,946,276.40**	
3	Conventional, BOW Method			
	a. 1 m2 Brick Installation 1/2 stone	80.85	69,064.03	5,583,826.83
	b. 1 m2 Plastering	161.7	24,725.20	3,998,064.84
	c. 1 m2 Edging	116.8	4,010.30	468,403.04
	Total cost		**10,050,294.71**	

From the table above it can be seen that the cost of wall work by using wall collapsing powder material is cheaper than cheaper wall work with conventional brick material with SNI and BOW methods. It should be noted, however, that the construction of walls with wall-ruins powder, with cast in place method requires additional tools such as cruher machine, mould, equipment, etc.

Table 2. The equipment for cast in place method (Cahyono, Eko Puji[6])

No	Tool type	Unit price (Rp)	amount	cost (Rp)
1	Stone Crusher	20.000.000,00	1 unit	20.000.000,00
2	Mould	1.719.550,00	11 set	18.915.050,00
3	Buffer	170.000,00	22 set	3.740.000,00
			Total cost	**42.655.050,00**

Procurement of equipment mentioned above, indirectly makes the unit price index for making the wall printing in place becomes more expensive. But the cost of procurement will be cheaper if it is charged to many homes. In other words, the more houses built, the price of the printed wall unit in place will be cheaper, near the price according to table 2. For more clearly can be seen in Figure 5 below.

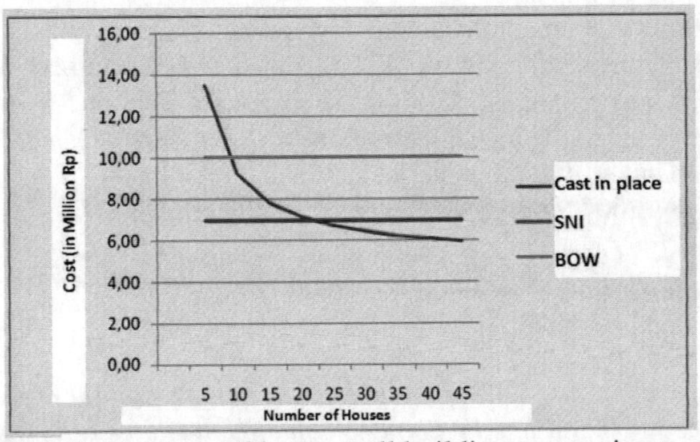

Figure 5. The Comparison of house wall building cost-estimate per m2, using cast in place method, SNI Analyzing Calculations, and BOW Analyzer Calculations (Cahyono, Eko Puji[6])

The cost of wall construction with cast in place method will be proportional to the work of the masonry pairs using the BOW method on the construction of the 7th house unit (± 10 million / m^2), whereas if compared to the SNI will experience the equivalence of the price on the construction of the 19th house unit (± 7 million / m^2), for more and more houses are built, the per-m2 price of the printed house walls will be cheaper.

In addition to cost factors, it turns out that cast in place method also has a predominantly in terms of time and work efficiency. This can be seen from the comparison table below.

Table 3. Comparison of the performance of cast in place methods with conventional methods (Yunus Anis, Mochamma[7])

Description	Cast in place method	Conventional method
Area of building	300 m^2	120 m^2
Work duration	70 days	60 days
Manpower	6 – 8 workers	14 – 18 workers

IOP Conf. Series: Materials Science and Engineering **401** (2018) 012017 doi:10.1088/1757-899X/401/1/012017

4. Conclusions

Recycled building materials from wall ruins can be used in powder form as a natural sand (fine aggegate) through crushing/recycling process. Mortar made from wall-ruins powder is suitable for use as a simple wall-construction material especially in mass construction. Fine aggregate from wall-ruins powder (through recycling material process) is more efficient and effective than natural sand if used for mass construction (more than 20 houses) at the same time. Work methods with cast on-site method also have an advantage in terms of time and work efficiency.

Referencess

[1] World Commission on Environment and Development 1987 *Our Common Future* (Oxford University Press)

[2] Roodman, D.M. & Lenssen, M 1995 *A Building Revolution: How Ecology and Health Concerns Are Transforming Construction* (Worldwatch Paper #124)

[3] Satyarno, I. 2007 *Pembuatan Kembali Rumah Tembok yang Runtuh Pasca Gempa dengan Sistem Daur Ulang Cetak di Tempat* (Seminar Nasional Teknologi Tepat Guna Penanganan Sarana Presarana di Indonesia)

[4] Satyarno, I. 2006 *Recycling Procedure of Brick Masonry Wall Rubble* (Proc. of the International Seminar and Symposium, Earthquake Engineering and Infrastructure & Building Retrofitting, August 28th 2006), Graduate School for Infrastructure Management Gadjah Mada University in Collaboration with Research Center For Engineering Science Gadjah Mada University, Yogyakarta.

[5] Tjokrodimulyo, K. 2007 *Teknologi Beton* (Biro Penerbit KMTS Fakultas Teknik Universitas Gadja Mada)

[6] Puji Cahyono, Eko 2007, *Perbandingan Rencana Anggaran Biaya Bangunan Konvensional dan Bangunan Cetak di Tempat dengan Analisa Metode SNI 2002* (Tugas Akhir JTSL FT UGM, Yogyakarta)

[7] Yunus Anis, Mochammad 2007 *Perbandingan Pelaksanaan Pembangunan Rumah Metode Cetak di Tempat dengan Pembangunan Rumah Konvensional* (Tugas Akhir JTSL FT UGM, Yogyakarta)

IOP Conf. Series: Materials Science and Engineering **401** (2018) 012018 doi:10.1088/1757-899X/401/1/012018

Batch biodegradability study of fresh and aged food waste: performance and kinetic

Roslinda Seswoya*, Muhammad Fahmi Abd Rahman and Nur Aiza Darnak

Micro Pollutant Research Centre (MPRC)
Faculty of Civil and Environmental Engineering,
Universiti Tun Hussein Onn Malaysia, 86400 Parit Raja, Johor, MALAYSIA

*roslinda@uthm.edu.my

Abstract. Peninsular Malaysia generates 17000 tonnes of municipal solid waste (MSW) per day in 2002 and it is estimated to increase to 31,000 tonnes per day on 2020, unfortunately food waste (FW) is the largest component of MSW. The high amount of FW generated related to landfill's main issues such as foul odour, toxic leachate, emission of greenhouse gases and vermin infestation. FW is highly potential in producing methane because it is rich in volatile solid. Therefore, the research trend on anaerobic digestion of FW has shifted for methane production because methane can be transformed to electrical energy production. The objective of this research is to evaluate the methane potential and kinetics of methane production from the anaerobic digestion of fresh and aged food waste taken from fast food restaurant. The methane production under mesophilic condition was monitored by using Automatic Methane Potential Test System II (AMPTS II). Triplicate sample reactors were prepared for each substrate (fresh and aged FW) at IS ratio of 2.0 on VS basis. Besides that, duplicate blank reactors were prepared concurrently, contained only inoculum. The fresh FW resulted in higher ultimate methane yield (M_o) and methane production rate (R_m). Comparing to results from fresh FW, the M_o and R_m observed from digestion of aged FW were decreased by 17% and 27% respectively. Kinetic analysis using Modified Gompertz modelling showed that M_o for fresh and aged FW were increased, as well as for lag phase (λ). Meanwhile the decreased R_m was observed for fresh and aged FW.

1. Introduction

Anaerobic digestion (AD) has been proven to be an efficient and green technology in disposing of sewage sludge, crop residues, food waste and animal manure [1], [2]. Anaerobic digestion is the condition in which several consortia of microorganisms simultaneously break down the biodegradable material - (organic wastes) without the presence of oxygen. Anaerobic digestion can be used to treat various organic wastes (substrate) and recover bio-energy in the form of biogas, which consists mainly of CH_4 and CO_2 [2]. The design and the performance of anaerobic digestion processes are affected by many factors such feedstock characteristics, reactor design and operation conditions. The physical and chemical characteristics of the organic waste are important information for designing and operating anaerobic digesters because they affect biogas production and process stability during anaerobic digestion [3].

A biochemical methane potential (BMP) test provides a preliminary indication of the biodegradability of a substrate (targeted organic waste) and its potential to produce methane via anaerobic digestion. The

Content from this work may be used under the terms of the Creative Commons Attribution 3.0 licence. Any further distribution of this work must maintain attribution to the author(s) and the title of the work, journal citation and DOI.
Published under licence by IOP Publishing Ltd

manually BMP setup requires periodically checking for the methane production by manual sampling and analysing the gas composition using gas chromatography (GC). Unfortunately, this approach is time- consuming and the regular checking should be at the earlier days of the BMP assay. Beside, the use of gas chromatography for the biogas composition identification stage is challenging due the method development. In practice, BMP test employed the large number of reactors and long incubation periods (e.g. over one month). Lately, the BMP test was conducted using the automatic methane potential test system (AMPTS) for on-line measurements of biodegradability tests. AMPTS II was used in anaerobic biodegradability study of FW [4], [5]. The inoculum to substrate (I/S) ratio of 2.0 was recommended for the BMP test for the digestion of FW in order to limit inhibition effects associated with accumulation of intermediate compounds, such as VFAs [6].

The Modified Gompertz modelling was typically applied model for the determination of methane production kinetics for the degradation of food waste [6]. The ultimate methane yield (M_o) of digestion FW observed from laboratory and Gompertz modelling differed slightly [6].

Based on literature search, it was found that no study has yet been conducted on the comparison of kinetic of methane production from the digestion of fresh FW and aged FW taken from fast food restaurant (generated in Malaysia). Commonly, the anaerobic digestibility study was conducted either for FW from university, canteen or household. Therefore, the main objective of this study is to evaluate the methane potential and kinetics of the digestion of FW collected from fast food restaurant. A series of batch BMP tests were conducted to investigate the effect of the storage time.

2. Material and methods

2.1. Substrate and inoculum

The FW was taken from the restaurant serving fast food. The FW was crushed by using kitchen blender for homogenizing the substrate [7]. The tap water is used for dilution [4]. In this study the fresh FW and aged FW were used as substrate, each at FW to tap water (FW:TP) ratios of 1:1 and 1:1.5 respectively. The inoculum, which was anaerobic bacteria were taken from existing full-scale anaerobic digester treating POME. It is a common practice to use anaerobic bacteria from existing anaerobic digester for the anaerobic digestibility study as described by Rajagopal [7]. The substrate and inoculum were stored in 4°C in refrigerator until use [7].

2.2. Experimental Setup

The batch biodegradability test or BMP tests were conducted using 500 mL of Duran bottle. The BMP assays were conducted using Automatic Methane Potential Test System (AMPTS 11) (Figure 1), in which only methane pass through to the gas volume measuring device and recorded in unit C [8]. Two series of BMP test were conducted; 1) fresh FW (substrate and inoculum were fresh) and 2) aged FW (substrate and inoculum were stored for 7 days in refrigerator at 4°C). The mass of substrate (FW) and the inoculum was calculated on the VS basis at inoculum to substrate ratio of 2.0. The reactors were purged with pure nitrogen in headspace to maintain anaerobic digestion condition were sealed [7]. The tests were done at mesophilic condition (37°C) [4] and agitated at 150 rpm [6]. The triplicate sample reactors (substrate with inoculum) and duplicate blank reactors (only inoculum) were prepared for each BMP experiment [4]. In the termination day of BMP assay, the pH of each reactor were measured to confirm the pH remained in the acceptable range for stable anaerobic process.

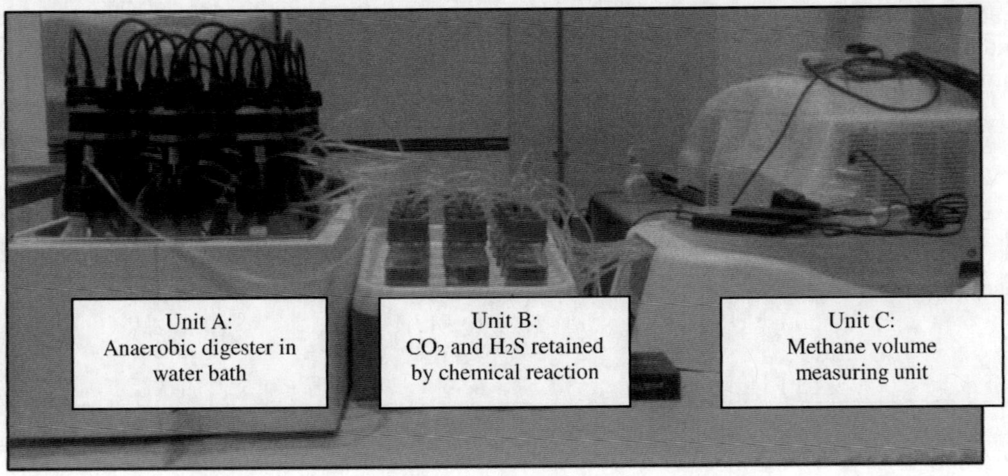

Figure 1. AMPTS II

2.3. Analytical Methods
The samples were measured for solids in g/L (TS and VS) and all tests were based on Standard Methods: procedure 2540G [9]. Meanwhile, VS in % was calculated following the calculation as shown by Bioprocess AB [8].

2.4. Batch kinetics modeling
In this study, the Modified Gompertz model was fitted to the experimentally observed cumulative methane production curves. The Modified Gompertz model has been widely applied in modelling batch methane production from FW digestion (Equation 1) [6], [10]. The parameter to be determined are ultimate methane yield (M_o) and maximum methane production rate (R_m) and lag phase (λ). Excel Solver was used to estimate M_o, R_m, and λ from the nonlinear graphs analysis which fit the experimental data set as recommended by Eskicioglu [11].

$$M = M_0 . \exp\left\{- \exp\left[\frac{R_m . e}{M_0}(\lambda - t) + 1\right]\right\} \quad (1)$$

Where,

M = Cumulative methane yield (mL /g VS added)
Mo = Ultimate methane yield (mL /g VS added)
Rm = Maximum methane production rate (mL /g VS added/day)
t = digestion time (t)
e = 2.718 [12]
λ = The lag phase time (day)

3. Results and Discussion

3.1. Substrate characteristics
The characteristics of substrates for this study are tabulated in Table 1. There was a difference of TS and VS between both samples, which fresh FW have higher TS and VS value compared to aged FW. These may occur because of the different dilution ratio used in the study. VS to TS ratios (VS/TS)

typically range from 0.80 to 0.97 [13]. Meanwhile, VS/TS ratio for FW taken from fast food restaurant located at France was 0.93 [14]. This study suggested that the storage time (less than 10 day) is insignificant in changing the VS/TS ratio.

Table 1. Characteristics of FW analysed in this study (N =3)

Parameter	Types of substrates	
	Fresh FW	Aged FW
Total solids, TS (g/L)	155.33 ± 20.28	94.22 ± 3.67
Volatile solid, VS (g/L)	150.56 ± 17.19	90.89 ± 2.69
Volatile solid, VS (%)	4.19 ± 0.42	2.59 ± 0.05
VS/TS Ratio	0.97	0.96

3.2. Ultimate methane yield

Figure 2 showed the comparison of methane yield curve of two samples. The ultimate methane yield from fresh FW is slightly higher than that what been observed from aged FW sample, each at 2349.3 ml CH_4/ g VS and 1933.0 ml CH_4/ g VS respectively. The ultimate methane yield observed from this study was higher; about 4.5 and 3.7 times more than what was observed from fast food restaurant as observed by Capson-Tojo [14]. The effect of storage time of the substrate also being studied in this study. This study suggested that the longer storage time resulted in the lower ultimate methane yield. Lü [15] observed the vice-versa condition, in which the longer storage time improved acidification efficiency, and provide better substrate for methanization. This subsequently leads to significantly increased methane production. The methane production from the aged FW (collected for canteen and being stored for 7 days) was about 1.8 to 2.4 times higher than the fresh FW.

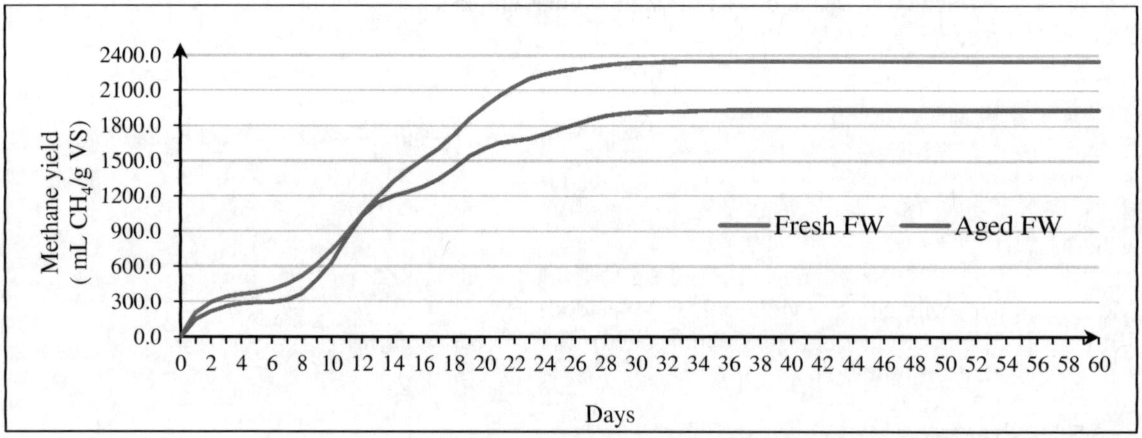

Figure 2. Methane yield curves from fresh and aged FW

3.3. Kinetics of methane production

The methane yield curve plotted using the laboratory data and Modified Gompertz modelling showed that more than 90% of the ultimate methane yield from fresh FW and aged FW were achieved at day 30. Therefore, as for the designing the anaerobic digester, the retention time could be assumed as 30 days. Table 2 showed the kinetic parameters observed from laboratory data and modelling. The ultimate methane yield (M_o) observed from modelling analysis for each digestion were higher than what was observed from the laboratory data, increased up to 7%. Meanwhile, the significant increase of lag phase (λ) and decrease of methane production rate (R_m) were observed from the modelling analysis. However, the lag phase (λ) is still less than two days.

Table 2. Kinetic equation parameters of methane production

Kinetic equation parameter	Fresh FW (Lab)	Fresh FW (Gompertz)	Aged FW (Lab)	Aged FW (Gompertz)
Mo (mL CH_4/gVS)	2349.35	2514.8	1933.02	2059.39
R_m (mL CH_4/gV/day)	148.8	93.24	108.08	79.63
λ (day)	0.04	0.98	0.04	1.58

4. Conclusion

The methane potential from the batch digestion of fresh and aged FW collected from fast food restaurant at inoculum to substrate (I/S) ratio of 2 under mesophilic condition showed a significant difference between them. However, the highest ultimate methane yield (M_o) and methane production rate (R_m) were observed from fresh FW. This showed that the digestion of this type of waste should be done freshly.

Acknowledgements

The authors wished to thank Universiti Tun Hussein Onn Malaysia for the funding of this research (TIER 1-Vot U844). We also wish to thank Cenergi SEA Sdn. Bhd for their cooperation

References

[1] X. Dai, N. Duan, B. Dong, and L. Dai, "High-solids anaerobic co-digestion of sewage sludge and food waste in comparison with mono digestions : Stability and performance," *Waste Manag.*, vol. 33, no. 2, pp. 308–316, 2013.

[2] M. S. Prabhu and S. Mutnuri, "Anaerobic co-digestion of sewage sludge and food waste," *Waste Manag. Res.*, vol. 34, no. 4, pp. 307–315, 2016.

[3] R. Zhang *et al.*, "Characterization of food waste as feedstock for anaerobic digestion," *Bioresour. Technol.*, vol. 98, pp. 929–935, 2007.

[4] V. Cabbai, M. Ballico, E. Aneggi, and D. Goi, "BMP tests of source selected OFMSW to evaluate anaerobic codigestion with sewage sludge," *Waste Manag.*, vol. 33, no. 7, pp. 1626–1632, 2013.

[5] Y. Li, Y. Jin, H. Li, A. Borrion, Z. Yu, and J. Li, "Kinetic studies on organic degradation and its impacts on improving methane production during anaerobic digestion of food waste," *Appl. Energy*, vol. 213, no. September 2017, pp. 136–147, 2018.

[6] G. De Gioannis, A. Muntoni, A. Polettini, R. Pomi, and D. Spiga, "Energy recovery from one- and two-stage anaerobic digestion of food waste," *Waste Manag.*, vol. 68, pp. 595–602, 2017.

[7] R. Rajagopal, J. W. Lim, Y. Mao, C.-L. Chen, and J. Y. Wang, "Anaerobic co-digestion of source

segregated brown water (feces-without-urine) and food waste: For Singapore context," *Sci. Total Environ.*, vol. 443, pp. 877–886, 2013.

[8] Bioprocess Control, *AMPTS II Operation and Maintenance Manual*. Sweden, 2014.

[9] APHA, *Standard Methods for the Examination of Water and Wastewater*, 20th ed. New York: Water Environmental Federation, 2005.

[10] B. Deepanraj, V. Sivasubramanian, and S. Jayaraj, "Effect of substrate pretreatment on biogas production through anaerobic digestion of food waste," *Int. J. Hydrogen Energy*, vol. 42, no. 42, pp. 26522–26528, 2017.

[11] C. Eskicioglu, K. J. Kennedy, and R. L. Droste, "Characterization of soluble organic matter of waste activated sludge before and after thermal pretreatment.," *Water Res.*, vol. 40, no. 20, pp. 3725–36, Dec. 2006.

[12] J. Li, S. M. Zicari, Z. Cui, and R. Zhang, "Processing anaerobic sludge for extended storage as anaerobic digester inoculum.," *Bioresour. Technol.*, vol. 166, pp. 201–10, Aug. 2014.

[13] Y. Li, Y. Jin, A. Borrion, H. Li, and J. Li, "Effects of organic composition on mesophilic anaerobic digestion of food waste," *Bioresour. Technol.*, vol. 244, pp. 213–224, 2017.

[14] G. Capson-Tojo *et al.*, "Accumulation of propionic acid during consecutive batch anaerobic digestion of commercial food waste," *Bioresour. Technol.*, vol. 245, pp. 724–733, 2017.

[15] F. Lü, X. Xu, L. Shao, and P. He, "Importance of storage time in mesophilic anaerobic digestion of food waste," *J. Environ. Sci.*, vol. 45, pp. 76–83, 2016.

IOP Conf. Series: Materials Science and Engineering **401** (2018) 012019 doi:10.1088/1757-899X/401/1/012019

Character of Traditional Street: An Overview of Physical Components Associated with Building, Landscape and Street Pattern in Royal Town

Wan Norisma Wan Ismail[1,2], Nor Haslina Ja'afar[2], Fadli Arabi[1] and Elina Mohd Husini[1]
[1] Department of Architecture, Faculty of Engineering and Built Environment, Universiti Sains Islam Malaysia
[2] Centre of Architecture and Built Environment Innovative (SeRAMBI), Programme Architecture, Secretariat of Identity of National Architecture, Faculty of Engineering and Built Environment, Universiti Kebangsaan Malaysia

norisma@usim.edu.my

Abstract. Streets are the most ubiquitous public spaces, serving a multitude of functions which tied to an extremely narrow range of historical, geographical and cultural settings. Streets record and determine the history of city form, more than any other element of the urban infrastructure. It has attributors in making it a successful street with its own unique identity. However, in the present situation these attributes are almost eroded by the uncontrolled development and modernity. Streets in Malaysia's royal town are no exception from this issue. This paper attempts to review the attributes of the physical components in contributing to street character of Kuala Kangsar, the royal town of Perak, Malaysia. The study covers a thorough literature review of physical components and its elements to determine the most important components captured. From this in-depth review, the study managed to identify three prominent physical components which strongly influenced the physical character of street in royal town. The study reveals that building, landscape and street pattern are the most extrusive components among others. These physical components will create physical qualities such as attractiveness, visibility, permeability, legibility, human scale and safety in creating street characters that symbolize the historical and cultural context. Therefore, establishing these attributes help to preserve the unique and invaluable historical heritage that has become the identity of the place.

1.0 Introduction

History of a town is written in its fabric. Streets are the first significant element which forms the image of a town. It is a physical manifestation of the public realm whereby it forms an interaction of human population with the architectural fabric. It is defined as one of the essential elements that acts as a public open space through its function as a communication channel and for social and activity space (Moughtin 2006; Krier 1979). Tibbalds (2001) highlighted that it is very easy for a city to lose its heritage. Streets record and determine the history of city form, more than any other element of the urban infrastructure (Moudon 1987). Mehta (2013) stated that in the infrastructure of a town, physical element of the street is the most essential element that forms the history and civilization of a place. The study of such physical elements can be seen from various perspectives either literally or metaphorically because the street is influenced by social, economic and political culture in developing an environment of high social value and sustainable living. Physical elemental aspects have been selected as the scope of the study because based on Ja'afar (2014); A Bashri & Shuhana (2012); Mulyadi (2008) and Shuhana (1997), physical

Content from this work may be used under the terms of the Creative Commons Attribution 3.0 licence. Any further distribution of this work must maintain attribution to the author(s) and the title of the work, journal citation and DOI.
Published under licence by IOP Publishing Ltd

elements have a significant impact on the character of the streets as well as creating sense of place. Thus, the scope of the study is to review the attributes of the main physical components in contributing to street character of the royal town of Kuala Kangsar. The study covers a thorough literature review of physical components and its elements to determine the most prominent components captured. The first part of this paper outlines the historical development of Kuala Kangsar, Perak as the case study. While the second part focuses on scientific findings which stipulates that buildings, landscapes and street patterns are the major components that contribute to a successful street.

2.0 The Royal Town of Kuala Kangsar

Kuala Kangsar is the royal town of Perak, Malaysia located at the downstream of Kangsar River, where it flows into the Perak River. According to Sodangi et al. (2013) from Heritage Trust of Malaysia (1990), Kuala Kangsar has been the home to Perak's royal family since Sultan Muzafar's reign in the 18th century. In the context of this town's history, physical components such as buildings, landscapes and street patterns have significant significance to street characters in this royal town. The town inherited rich legacy of heritage buildings with unique architectural and historical values such as royal palaces, colonial buildings, mosques, traditional Malay houses and monuments which act as landmarks to reflect its character, sense of place and place attachment to the local and foreign visitors. Istana Kenangan, Istana Iskandariah, Sultan Azlan Shah Museum, Ubudiah Mosque, Malay College, Kuala Kangsar Clock Tower are among the examples of buildings named. From the landscape aspects, the town is the site of the first rubber tree planted in Malaysia by an English botanist Henry Nicholas Ridley. The waters of Sungai Perak, which flow through the town has become one of the attractions for its activities and tourism industry thus creating a sense of place through its relationship between people and its atmosphere. Street pattern are narrow and pleasant and great for a stroll as it winds round the grassy hill of Bukit Chandan, a majestic hill where the royal palaces are located. Today, the town which holds historical values, still holds its old world's charms - preserved colonial buildings, streets with century-old shady trees as well as well-landscaped gardens surrounding the royal enclave of Kuala Kangsar. It is a royal town intertwined with some colonial heritage and spectacular natural sceneries.

3.0 Problem Statement

The existence of royal towns in Malaysia based on a study by NA Samsudin et al. (2018), affirmed that these heritage areas (including the streets) will be diminishing throughout time due to improper planning and modernization of the royal towns if there are no actions taken to protect its identity and characters. N Zalina et al., (2015) highlighted that industrialization has caused the growth of urbanization within the area of royal town, affecting valuable elements of its urban heritage. While part of the city fabric along heritage streets in Johor Bahru were already torn down to meet the demands of the modern society. Action by local authorities are being taken to preserve the identity and character of streets and other places that are shaped by the physical elements, image and activities (Ismail, W. H. W, 2015). Another royal town in Alor Setar, Kedah according to SM Shukri et al., (2017) experienced rapid development and urbanization causing irreversible transformation of their form and character and perhaps in turn, with loss of historic identity and lose their function in meeting people's needs towards the places and the streets connecting it. The findings show that poor connectivity among the physical elements had influenced the identity of the royal town itself.

According to Fauziah et al (2014), cities with historical heritage such as Kuala Kangsar need to emphasize the importance of place identity and reinforce its local uniqueness to make the appearance of the town and streets more distinctive. Study by R Hashim (2001) mentioned that another royal town at Pekan, Pahang has experienced both rapid economic growth and physical development which affects the royal townscape including its streets. Result shows that many buildings have been demolished and replaced with new buildings that lack of local characteristics. Consequently, the new development has become a threat to the preservation of the local cultural heritage values including its heritage streets. Based on the issues outlined above, the scenario had threatened the character of town and the streets

IOP Conf. Series: Materials Science and Engineering **401** (2018) 012019 doi:10.1088/1757-899X/401/1/012019

surrounding it resulting in the loss of street character in a royal town. Hence it is very crutial to establish the physical components in shaping the street character, identity and meaning of towns.

4.0 Theoretical Framework Based on Street Character Theory

Based on a study by Ja'afar (2014) and Musaab & Rawia (2017), street is divided into two, namely physical appearance and function. Street appearance refers to buildings, landscapes, street patterns, quality of view, location, street structure, movement patterns, parking, street widths, street network patterns and sensory (Nor Haslina 2014; Abbaszadeh 2011; Lindsay et al 2009; Moughtin & Martens 2006; Oktay 2005; Lillebye 2001; Appleyard 1980). While function refers to the activities that occur around it because the activity contributes to the success of the street thus creating the atmosphere of the place (Ja'afar 2014; Ja'afar 2011; Gehl 2010; Shuhana 1997). Hence, this study will discuss the physical components from both aspects. Based on the detailed review on the elements and attributes that contribute to the design and the character of street, a theoretical framework has been developed to enable this lengthy description to be viewed graphically (refer Figure 1). It can be concluded that the main components that contribute to the physical character of the streets are buildings, landscapes and street patterns.

Figure 1: Theoretical Framework

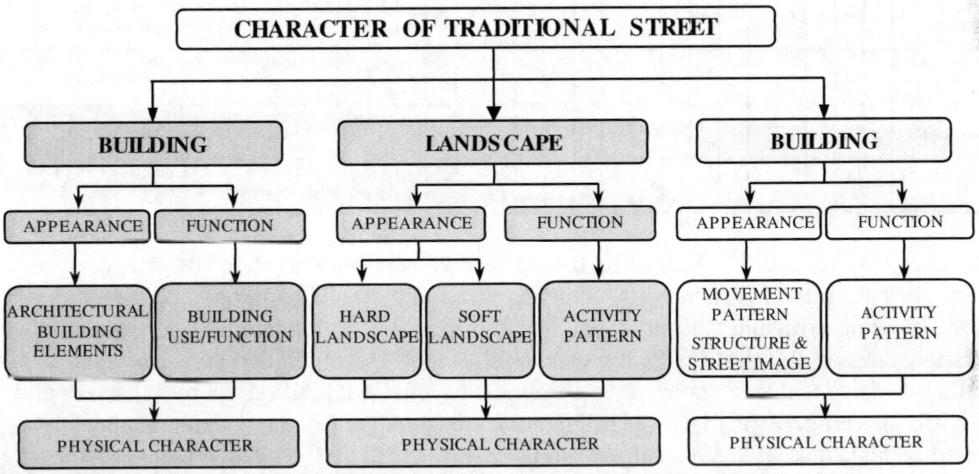

5.0 Physical Components Associated with Street Character of Royal Town

In the urban design field, there are four fundamental elements that make up an environment namely physical, behavioral, psychological and social (A Bashri, 2012) This paper focuses only on the physical aspects of the street because according to Musaab & Rawia (2017), Ja'afar (2014) and Lindsay (2009), streets or paths are the basic things that are closely related to the character of a place. Good street design is very important in creating a successful street. Successful street design is a summary of the physical environment formed from the relationship of the various components and elements in which this relationship intertwined with each other to produce a unique street character (Ja'afar 2014; Lindsay 2009; Ziezel 1984; Manadipour 1996, Rapoport 1977). Hence, these physical characters directly linked to genius loci, identity, meaning and symbols for society (Jaafar, 2014; Liang 2010; Oktay 2005; A Bashri & Shuhana 2008. 2005; Thomas 2002; Kropf 1996).

5.1 Building as Physical Component - Building is the most dominant physical component and is the major contributor to shaping street characters. Buildings and streets are two essential elements that have a significant impact on the image of a town (Amirhossein et al. 2013) and these buildings affected the street character through its appearance and function (Nor Haslina 2014; Moughtin & Martens, 2006; Alfrey, 2010).

IOP Conf. Series: Materials Science and Engineering **401** (2018) 012019 doi:10.1088/1757-899X/401/1/012019

Figure 2: Summary of physical element and physical character for building component

BUILDING COMPONENTS																					
		APPEARANCE															FUNCTION				
		ARCHITECTURAL BUILDING ELEMENTS															BUILDING USE/FUNCTION				
	PHYSICAL ELEMENT / PHYSICAL CHARACTER	Location of Building	Building Function	Building Age	Building Size	Building Height	Building Structure	Condition of Building	Building Setback	Building Facade	Façade Opening	Architectural Style	Roof Design	Façade Modification	Building Signage	Quality of View	Mixed-used Buildings	Commercial Building	Civic Building	Residential Building	REFERENCES
APPEARANCE	Safety & Comfort	•	•	•		•		•	•		•	•			•	•					Musaab & Rawia, 2017; R. Oppong et al. 2017; N. Rahman 2016; Norhafizah et al. 2014, Nor Haslina 2014; Sharifah Salwa et al. 2015; W. Hashimah 2014; Askari et al 2013; Baper & Hassan 2012; Abbaszadeh 2011; Askari & Dola 2009; CABE 2008; Cooper 2005; Carmona et al. 2003; NYC DOT 2009; Shirvani 1985; Jacobs 1996; Carr et al. 1992, Alexander 1977
	Attractiveness	•								•	•	•	•	•		•					Nor Haslina 2014; Sharifah Salwa 2013; Moughtin & Martens 2006; A. Abbasiasbagh 2013; Le Ann 2006; Matthew 2003
	Human Scale		•		•	•				•		•									Drielle & David 2015; Change Lab Solution 2012; Borgett 2011; Ewing & Handy 2009; City of Meridien 2009; Purciel et al. 2006, Lucu 1995; Jacobs 1993
	Legibility & Visibility									•	•	•	•	•	•						N. Ginting et al. 2016; K. Nakamura 2016; Drielle & David 2015;M.Mahmoudi et al. 2014; Wagner & Caves 2012; Abazzadeh 2011; Tan & Gehl 2006; Gehl et al. 2006; Gehl 1987; Venturi 1986; Appleyard 1980;
	Continuity & Consistency		•	•	•	•	•	•	•	•	•	•	•	•	•						Ohadesh et al 2014; A.Abbasiasbagh et al. 2013; Baper & Hassan 2012; West Concorde Village Centre 2011; Le Ann 2006; Matthew 2003; Sharifah Salwa 2013; SZA Idid 2008; Moughtin & Martens 2006; Shirvani 1985
FUNCTION	Attractiveness																•	•	•		Nor Haslina 2014 & 2006; Sharifah Salwa et al. 2015; Shuhana 2011 & 1997; CABE 2008; Tibbalds 2001; Duany 2000; Bentley 1985; Jacobs 1969;
	Safety & Comfort																•			•	W. Hashimah 2014; Nor Haslina 2014 & 2006; Shuhana 2011 & 1997; Idid 2008; Shuhana & Norshidah 2008; Davies 2000; Francis 1987
	Accessibility Imageability Legibility																	•	•		Sharifah Salwa et al.2015; Nor Haslina 2014 & 2006; W Hashimah 2014; Shuhana 2011 & 1997; Norshidah 2010; Norshidah & Shuhana 2008; A Bashri et al. 2008

5.1.1 Building Appearance - Based on literature review, physical elements that derived from the appearance of building components are - building location, types of building, size, height, age, building condition, building structure, facade walls, facade openings, architecture style, roof design, façade modification, signage and quality of view (Nor Haslina 2014; Shuhana 2011; Abbaszadeh 2011). These physical appearance features characters such as attractiveness, legibility, visibility, safety and comfort (refer Figure 2). Venturi et al. (1986) highlighted that building appearance which includes design, form, facade and architectural style makes it identifiable, legible and visible. These elements should relate to each other in scale to create a viable urban streetscape create a consistent building edge that frames and activates the public realm. Moughtin & Martens (2006) stated that building facade can be used to analyze the aesthetic and urban design qualities as it is a physical element associated with the principle of architectural composition such as scale, rhythm, position, harmony, balance and unity. While openings on building facade attempts to produce rhythm and pattern that highlight the uniqueness of the building details.

To preserve the town and street characteristic, physical elements at the building façade such as doors, windows, arcades and balconies play a significant role in demonstrating architectural continuity (A. Abbasiasbagh, 2013, Matthew 2003). Building appearance is also influenced by the building-to-human scale factor, where human scale refers to the height of the building as opposed to the human eye view of the surrounding building with the right scale of view (Jacobs 1993). Building signage which contributes to pedestrian legibility are important in the consistency and continuity of overall building architecture (Shirvani 1985) which aimed at reducing the negative visual impact and avoiding confusion with traffic signs. Apart from that, other elements such as building skyline, roofscape, building shape and layout, details and articulation of facade, window and door orientation to the streets are elements of building appearance that creates unique quality of view (Ja'afar 2014; Baper & Hassan 2012).

5.1.2 Building Use/Function - Function and use of buildings and the streets in its surroundings are related to the type of economy/goods/services offered around it (Nor Haslina et al 2011; Shuhana et al., 2004). Reviews by Norhafizah (2014); Miskell (2008); Gehl (2000); Tibbalds (1992) proved that the types and functions of the building are the main factors affecting the physical character of a street and thus contribute to the success of a place. According to previous study, there are four categories of buildings based on their functions and uses: mixed used buildings, commercial, residential and civic buildings (Ja'afar 2014; Shuhana 2011; Abbaszadeh 2011). The functions and uses of the building play a key role in shaping local activities and human behaviors to create a sense of place or place attachment that makes it memorable (Appleyard, 1980). Functions for each type of building use will be examined in relation to the outdoor and indoor activities of the building as well as its impact on the lower floors and the upper floors of the building.

5.2 Landscape as Physical Component - The landscape component also affects the design and character of the street through two aspects - appearance and function. Landscapes have a broad meaning in which it encompasses the geological structure of the land, flora and fauna, the pattern of human activity or the interaction between natural resources and human needs that emphasize the beauty and aesthetic value, architecture and overall ecology and history of a place. According to Ahmad Fauzi (2015), landscapes are divided into two - soft landscape and hard landscape. Soft landscape covers a wide variety of plants found in public parks such as trees, shrubs, palms and ground covers. It plays a vital role in applying natural values and elements of nature into the built environment (Moughtin et al. 1995). Meanwhile hard landscape refers to all types of man-made structures such as buildings, public utilities, infrastructure, antiquities and disabled facilities (Hussain & Ahmad 2010)

Figure 3: Summary of physical element and physical character for landscape component

LANDSCAPE COMPONENTS													
		APPEARANCE										FUNCTION	
		SOFT LANDSCAPE					HARD LANDSCAPE						
	PHYSICAL ELEMENT / PHYSICAL CHARACTER	Location	Vegetation/ Plants	Geography & Topography	Water Elements	Quality of View	Location	Pavement	Pedestrian Walkway	Street Furniture	Quality of View	Soft Landscape Hard Landscape	REFERENCES
APPEARANCE SOFT LANDSCAPE	Attractiveness	•	•		•	•							Mohd Afendi 2016; Drielle & David 2015; Nor Haslina 2014; N Syala 2011; Shuhana 2011; Moughtin & Martens 2006; Ahamad 2004; Davies 2000; Jacobs 1993; Alexander 1977
	Visibility		•	•		•							Nor Haslina 2014; Shuhana et al 2013; Simkins 2012; Shuhana 2011; English Heritage 2010; Tibbalds 2001;
	Comfort		•		•	•							Mohd Afendi 2016; Drielle & David 2015; Norhafizah et al. 2014; Eaton 2006; Shuhana et al 2004; Carmona et al. 2003; City of Austin 2001; Carr et al. 1992; Jacobs 1993; Carr et al. 1992; Kaplan & Kaplan 1982
	Human Scale	•	•										Drielle & David 2015; Ewing et al 2006; Nor Haslina 2014 & 2012; Crankshaw 1996 & 2009;
	Enclosure	•	•	•									Drielle & David 2015; V. Mehta 2006; Handy et al 2002; Jacobs 1993; Carr et al. 1992;
APPEARANCE HARD LANDSCAPE	Attractiveness						•	•	•	•			Drielle & David 2015; Nor Haslina 2014;; English Heritage 2010; Crankshaw 2009; Shaftoe 2008, Duggan 2007; Moughtin et al. 1995;
	Visibility/ Accessibility						•	•	•	•	•		Nor Haslina 2014; A. Bashri 2012; Nor Haslina 2011; Mackett et al 2008; Anwar 2006; Voltolini 2006; Tyrell 2004; A Bashri & Zubir 2004; SZA.Idid 1995
	Comfort						•	•	•	•	•		Nor Haslina 2014; A. Bashri 2012; Nor Haslina 2011; Mackett et al 2008; Anwar 2006; Voltolini 2006; Tyrell 2004; A Bashri & Zubir 2004; City of Austin 2001; Mijan 2000; SZA.Idid 1995; Carr et al 1992;
	Safety						•	•	•	•	•		Nor Haslina 2014; A. Bashri 2012; Nia & Rosley 2012; Nor Haslina 2011; Lindsay 2009; Colcester Borough Council 2009; Mackett et al 2008; Duggan 2007; Anwar 2006; Voltolini 2006; Tyrell 2004; A Bashri & Zubir 2004; Davies 2000; SZA.Idid 1995; Appleyard 1990; Shirvani 1985; Cullen 1985
FUNCTION	Comfort											•	Nor Haslina 2014 & 2011, N Syala 2011; Norshidah 2008; V Mehta 2006; Anwar 2006; Moughtin & Martens 2006; Voltolini 2006; Tyrell 2004; A Bashri & Zubir 2004; Ahamad 2004; Jacobs 1993
	Safety											•	Nor Haslina 2014 & 2011; Shuhana 2011; Anwar 2006; Tyrell 2004; A. Bashri & Zubir 2004; Nia & Rosley 2012; Mijan 2000; Lindsay 2009; Carr et al 1992

5.2.1 Landscape Appearance - Soft landscapes that surround the natural environment such as hills, rivers and seas influence urban structures and patterns through its responsive builds on topographic profiles, therefore create sense of attractiveness and visibility (Shuhana 2011). Physical elements of soft landscape appearance are - types of vegetation, geographical and topographical, water element, location of landscape and quality of view. Attractiveness, visibility, human scale, comfort and enclosure are the characters generated by these elements. While hard landscape appearance is categorized into three main elements - pavement, pedestrian path and street furniture (Ja'afar 2014). Street pavement (involving textures, materials, design patterns, orientations and layout arrangements) are used to create comfort, legibility and visibility for street users. Street furniture which is divided into three types of amenities - pedestrian, traffic and utility, consists of elements for example bench, streetlight, bus stop, toilet etc. resulting in a comfortable and safe street environment. Besides that, signage, traffic sign, artificial light, pedestrian path and its compliance to the needs of the disabled are also important in ensuring safety, comfortability, visibility and accessibility.

5.2.2 Landscape Use/Function - The function of street is related to human interaction between building and its surrounding landscape (Ja'afar 2014; Shuhana 1997). Gehl (2010) notes that activity within the natural setting is one of the key factor in creating a successful street. A study by Nor Haslina (2014) and Shuhana (2011) concluded that the landscape component greatly influenced the character and activities held along the street. Rapoport (1977) uses landscape components as a variable for analyzing activities because activity on the street produces high legibility. Lynch (1960) considers that activities are divided into two - human activities and activities arising from mechanical movements such as vehicles. Therefore, the function of landscape component is viewed from the aspect of human activity or vehicle and the relationship between human behavior and the surrounding landscape (Shuhana 2011; A. Bashri 1988). According to Mehta (2006) and Jacobs (1993), human activities associated with the natural environment will form a high quality of transparency and provide a safe and comfortable street environment (refer Figure 3).

5.3 Street Pattern as Physical Components - Street pattern is the result of the combination of three main elements known as land, building and street. The diversity of designs and patterns of these elements has produced several types of street patterns such as linear, grid and others (Alfrey 2010). The study of street pattern is assessed in terms of the historical and cultural heritage impacts as it greatly influences the pattern of the street and the character of place. Shuhana (2011) and Marshall (2005) defines street pattern as a building block layout pattern that affects the space of a town. A street pattern is formed when all the patterns, characters and features are combined. This will then result into a street pattern enclaved within a balanced environment that have physical qualities such as accessibility, permeability, legibility and visibility (refer Figure 4).

5.3.1 Street Pattern Appearance - The appearance of street pattern is divided into several elements namely movement pattern, structure and street image, location and quality of view. The movement pattern on the street permeability is related to the quality of connectivity and its linkages. According to Ja'afar (2014), movement patterns are assessed based on permeability studies. Permeability is an element of physical quality that illustrates the extent to which an environment can allow one to pass through one place to another. To facilitate users to easily explore streetscape and townscape, connectivity and linkages are intertwined with these movement pattern elements (Shuhana 2011; Bentley et al. 1985). Priority sequences for types of movement patterns in public spaces are preceded by pedestrians, cyclists, public transport users and lastly private vehicles (Ja'afar 2014; CABE 2010). According to Lynch (1960), the image of a city including its street has a strong connection with the five main elements - path, edges, districts, nodes and landmarks, whereby these elements are a major contributor to legibility of the street.

5.3.2 Street Pattern Function – The function of street is largely related to human activity that interacts with a place as well as how buildings and landscapes are used (Shuhana 1997). Activity is behavior response through social, cultural, economic and weather aspects of their environment. Activity provides strength, enthusiasm and character on streets and open space because physical space and activity are two complementary elements. Activity and space are two elements of the urban design principles that are interconnected. The presence of this activity will create a specific function/activity through the location and character of place. The presence of these activities presents distinctiveness, environmental character and the quality of view of the town (Shuhana 2011; Anwar 2006). Lynch (1960) notes that the existence of special function and activities along the street pattern highlights the attraction of the place and able to give a strong memory to the observer. A study by Abazzadeh (2011, 2009) also concludes that special activity creates an element of attractiveness to encourage the presence of the public on the street. All these lead to accessibility, legibility and safety which contributes to the success of the place (Alvarez 2009; Norhafizah 2012; Ja'afar 2014).

Figure 4: Summary of physical element and physical character for street pattern component

STREET PATTERN COMPONENTS							
		APPEARANCE				FUNCTION	
	PHYSICAL ELEMENT / PHYSICAL CHARACTER	Location	Movement Pattern	Structure & Street Image (nodes, landmark, path, edges, district)	Quality of View	Types of Activities	REFERENCES
APPEARANCE	Permeability		•				Nor Haslina 2014 & 2006; ; Shuhana et al 2012; Walk San Diego 2011; Wallhead 2011; Borgett 2011; CABE 2010; Barrie 2010; New Heaven 2010; Davies 2000; Lindsay 2009; Crankshaw 2009; Lusher et al. 2008; Taylor 2006; NYBC 2002; Bentley et al. 1985; Shirvani 1985; Jacobs 1993; Halph 1972
	Legibility Visibility	•	•	•	•	•	M. Mahmoudi 2014; Nor Haslina 2014 & 2011; South Dublin City Council 2011; Shuhana 2011; Abbazzadeh 2011; English Heritage 2010; Lindsay et al 2009; Eslami & Ismail 2009; Taylor 2006; Shokouhi 2003; Watson et al 2003; Lynch 1960; Kropf 1996; Cullen 1985; Rapoport 1990; Worskett 1970
	Accessibility		•	•			Sharifah Salwa 2013; Bogert 2012; Shuhana 2011; Barrie 2010; Haslina 2006; Marshall 2005; Davies 2000
FUNCTION	Attractiveness					•	Nor Haslina 2014; Shuhana 2011; Abbazzadeh 2011 & 2009; Alvarez 2009; Mulyadi 2008; Anwar 2006; Lynch 1960
	Legibility					•	Nor Haslina 2014 & 2011; Shuhana 2011; Mulyadi 2008; Anwar 2006; Taylor 2006; Montgomery 1995, Rapoport 1990
	Safety					•	Sharifah Salwa et al 2015; Norhafizah 2012; PPS 2011; Selamat 2005

6.0 Conclusion

The findings of the study show that three prominent physical components known as buildings, landscapes and street patterns serve as benchmarks that need to be emphasized in establishing street characters of royal town. These three components are interconnected where each component covers the two main aspects of appearance and function. The presence of historic building structures and all the architectural building elements have contributed to the uniqueness and character of the street. The existence of soft landscape and hard landscape elements are capable of producing a distinctive quality of view, contributes to social interaction and comfortable atmosphere as well as highlighting the character of the town and street. While street pattern that emphasizes on movement pattern and the five physical elements namely nodes, paths, districts, landmarks and edges contribute to an identifiable, accessible and clear form of environment. These physical components will ultimately contribute to physical qualities of the streets namely attractiveness, legibility, visibility, permeability, human scale and special activities.

Acknowledgements

This work was supported in part by Universiti Kebangsaan Malaysia (UKM) under Grant No. FRGS/1/2015/SSI11/UKM/02/2

References

[1] Ahmad Bashri Sulaiman (2012). National Seminar on Sustainable Urban Design for Livable Cities (SUDLiC 2012); 23 May 2012. In Sustainable Cities by Design: Breaking the Silos.

[2] Ahmad Basri, S., & Suhanna, S. (2008). Lessons from Traditional Towns and Cities in Malaysia for Sustainable Future Places. Ahmad Basri, S. and Mouktar, MM Urban Design Issues in the Developing World: The Case of Malaysia and Nigeria, 1-21.

[3] Abaszadeh, F (2009). Characterization of Urban Environments: A Direction for Persian Future. Design Principles and Practices: An International Journal, 3(4) 41-56

[4] Abaszadeh, F (2011). Characterization of Streets Physical Elements in Iran. Thesis Degree of Doctor of Philosophy. Fakulti Alam Bina, Universiti Teknologi Malaysia.

[5] Anwar, M. (2006). Model Rekabentuk Bandar Berasaskan Rekabentuk Bandar Tradisional. Master Architecture. Fakulti Alam Bina, Universiti Teknologi Malaysia.

[6] Appleyard, D. (1980). Livable streets: protected neighborhoods? The Annals of the American Academy of Political and Social Science, 451(1), 106-117

[7] Gehl, J. (2010). Cities for People. Island Press.

[8] Heritage Trust of Malaysia. Malaysian Architectural Heritage Survey: A Handbook. Kuala Lumpur: Badan Warisan Malaysia, 1990

[9] Ismail, W.H.W. (2015). Contemporary Uses of Buildings on a Heritage Street. Procedia-Social and Behavioral Sciences, 170, 633-641.

[10] Jacob, A. 91993). The Great Streets. United States of America. The MIT Press.

[11] Krier Rob. (1979). Urban Space. New York: Rizolli.

[12] Lynch K. (1960). Image of the City. The MIT Press, Cambridge, Massachusetts and London, England.

[13] Marchall, S (2005) Urban Pattern Specification. Solutions (pp.1-48). Solutions Project.

[14] Mehta, V. (2013). The street: a quintessential social public space. Routledge.

[15] Mulyadi, L. (2008). Karakter dan Konsep Susunatur Ruang Bandar Bersejarah Cakranegara Indonesia. Thesis Degree of Doctor of Philosophy. Fakulti Alam Bina, Universiti Teknologi Malaysia.

[16] Moudon, A. V. (1987). Public Streets for Public Use.

[17] Moughtin, J. C. (2006). Urban Design: Green Dimensions. Routledge.

[18] Moughtin, C., Oc, T., & Tiesdell, S. (1985). Urban Design Ornament and Decoration. Butterworth Architecture

[19] Musaab Sami Al-Obeidy and Rawia Marwan Dabdoob. (2017); Reviewing On Street Attributes In Influencing Sense Of Place And Place Attachment. Int. J. of Adv. Res. 5 (2). 356-365] (ISSN 2320-5407)

[20] Nor Haslina Ja'afar (2014). Karakter Fizikal Jalan Tradisional, Kajian Kes di Melaka Malaysia. Thesis Degree of Doctor of Philosophy. Fakulti Alam Bina, Universiti Teknologi Malaysia.

[21] Nor Haslina Ja'afar, Ahmad Bashri Sulaiman & Shuhana Shamsuddin (2011). IJAS International Multidisciplinary Conferences. Traditional Street Activities in Kuala Lumpur City Centre.

[22] Harun, N. Z., Fairuz, D. N., & Nordin, N. A. (2015). The roles of urban heritage in determining the image of the royal town of Sri Menanti, Negeri Sembilan. PLANNING MALAYSIA JOURNAL, 13(5).

[23] Rapoport. A. (1990). History and Precedent in Environmental Design. Plenum Press New York, London.

[24] Sodangi, M., Khamidi, M. F., & Idrus, A. (2013). Maintenance Management Challenges for Heritage Buildings Used As Royal Museums in Malaysia. Journal of Applied Sciences and Environmental Sustainability.

[25] Shukri, S. M., Wahab, M. H., Ibrahim, M. A., & Rozaly, M. Z. (2017). Sd 26 - Physical Characteristics of a Public Square in Malay Royal Town. Sustainable and Resilient Cities: The Way Forward.

[26] Shamsuddin, S. (2011). Townscape Revisited; Unravelling the Character of the Historic townscape in Malaysia. UTM.
[27] Samsudin, N. A., Rosley, M. S. F., Shahminan, R. N. R., & Mohamad, S. (2018). Preserving the Characteristics of Urban Heritage: An Insight into the Concept of Malaysian Royal Towns. Environment-Behaviour Proceedings Journal, 3(7).
[28] Shamsuddin, S. (1997). Identity of place: a case study of Kuantan town centre, Malaysia (Doctoral dissertation, University of Nottingham).
[29] Shukri, S. M., Wahab, M. H., & Rozaly, M. A. I. M. Z. Physical Characteristics of A Public Square In Malay Royal Town.
[30] Shirvani, H (1985). The Urban Design Process. Van Nostrand Reinhold Company, New York.
[31] Tibbalds, F. (Ed.). (2012). Making people-friendly towns: Improving the public environment in towns and cities. Taylor & Francis.
[32] Trancik, R (1986). Finding Lost Space; Theories of Urban Design, New York. Van Nostrand Reinhold.

Physical and Safety Features Of Ablution Spaces in the Mosques of Selangor and Kuala Lumpur

Julaila Abdul Rahman1, Zumahiran Kamarudin[1], Mohammad Hariri Abdullah[2], Ismail Jasmani[1], Nurfadhila Ramli[1]

[1]Kulliyyah of Architecture and Environmental Design, International Islamic University Malaysi, Gombak Campus, P.O Box 10, 50728 Kuala Lumpur, Malaysia

[2]Malaysia Design Council, Level 11 & 12 West Wing, Menara MATRADE, Jalan Khidmat Usaha Off Jalan Duta, 50480, Wilayah Persekutuan Kuala Lumpur, Malaysia

julailarahman@iium.edu.my

Abstract. This study aims to explore the condition of physical and safety features of existing ablution platform and spaces of selected Malaysian mosques which are located in Kuala Lumpur and Selangor. Therefore, the objective of this study was to analyze the physical condition of ablution platforms and spaces focusing on the ergonomic aspects, user-friendly and safety features. Direct observation was conducted at the 15 selected mosques to examine the design aspects of the ablution platforms and spaces. The focus of analysis on the physical elements of the ablutions including design forms; ergonomic aspects, and safety features. The study found that most of the mosques have provided ablution seat with an ergonomic aspect; the appropriate size of seats and allocation of the water faucet. However, most of the mosques have their ablution areas with inappropriate distance among the seats and platform height. In addition, there is a lack of design consideration towards the needs of PwDs (Person with Disability) in providing the ablution platforms. These include the ablution area are not easily accessible and there is no special seats or area provided for PwDs to perform their ablution. This study recommends improving the ablution platform and spaces to be more user-friendly with ergonomic design and safety features to ensure the smoothness of ablution activity.
KEYWORDS: Ablution platform, physical elements, mosques

1. INTRODUCTION

The Mosque is a focal point and most frequented place for a Muslim, after home and workplace. The Mosque primarily is a place of worship of Allah in 'jammah' and also serve more than as congregational prayer places five times daily. Apart from these, the mosque plays various important roles which include the community centre, learning centre, the seat of government, the welfare centre, prison, hospital and legitimate recreational activities which are stated in the holy Quran and hadiths. All Muslim generations go to mosques, regardless of their races and physical condition; normal and handicapped. The Muslim community has the benefit of using the facilities at the mosqueas one 'ummah'. (Jasmani Ismail et. Al,2017). There are many facilities provided in a mosque to cater the various needs and demands of Muslim users. One of the facilities that are required in a mosque is the ablution area. Muslim users use the ablution area, which consists of ablution platform including water faucet and other spaces for the comfortability of user to take their ablution ritual before prayer.

Content from this work may be used under the terms of the Creative Commons Attribution 3.0 licence. Any further distribution of this work must maintain attribution to the author(s) and the title of the work, journal citation and DOI.
Published under licence by IOP Publishing Ltd

IOP Conf. Series: Materials Science and Engineering **401** (2018) 012020 doi:10.1088/1757-899X/401/1/012020

In Malaysia, there are almost 6, 233 mosques were built throughout 14 states and there are many ablution areas were developed. There are very few studies have been done to check whether the ablution platforms and its physical conditions meet the user's needs and demands during the ablution activity. Past study showed that many ablution platforms in mosques are not user-friendly and ergonomic. This condition should not be overlooked because Muslims use the ablution platforms frequently, 5 times a day or more. It is essential to develop the ablution platform and its area towards ergonomic, user-friendly and practical needs so that many users can take their ablution with the comfortable and safe condition. The aim of this study is to explore the condition of physical elements of existing ablution platform and spaces found in Malaysian mosques located in Kuala Lumpur and Selangor areas. Therefore, the objective of this study was to analyze the physical condition of ablution platforms and spaces focusing on the ergonomic aspects, user-friendly and safety features.

2. BACKGROUND STUDIES

2.1. Mosques in Malaysia

In Malaysia, there are 2 kinds of place, namely mosques and surau where Muslims can perform their worship of Allah. A mosque usually has a bigger size with many spaces than 'surau'. In addition, 'surau' is used to refer to a prayer room in public buildings such as in shopping malls, offices, and others. A total of 6, 233 mosques and 18, 019 'surau' were built throughout14 states in Malaysia. The mosques in Malaysia are categorized into 4 types; national, state, district and village. There are 3 nationally prominent mosques in Malaysia, namely Masjid Putra, Masjid Negara and Masjid Tuanku Mizan Zainal Abidin. These mosques are located in Wilayah Persekutuan Putrajaya and Wilayah Persekutuan Kuala Lumpur. All mosques under this category are the official places for distinguished people and guests to perform their prayer. Most of the states in Malaysia have at least 1 mosque under state category. There are many mosques under district and village categories throughout the country. Each category of the mosque has different features including design form and building size. For example, the district mosque can at least accommodate up to 1,000 worshippers at any one time. Masjid Sultan Salahudin Abdul Aziz Shah (blue mosque) in Shah Alam, Selangor is the country's largest state mosque can accommodate up to 24,000 worshippers at any one time. The architecture of state mosques has a blend of vernacular, Moorish, Islamic architecture and modern style (A. Ghafar, 1999).

2.2 Conventional Ablution Platform Design

According to Mokhtar (2005), there are several possible models for the design of an ablution unit (refer table 1). All models have been designed with an anti-slip tile to ensure user safety and comfortability while performing their ablution. The Model 1 is the simplest design that includes a shelf for users to put their belongings and as a support of balancing their bodies by gripping strictly on the shelf. Model 2 shows an ablution with a seat. Most of the mosques in Malaysia adopted this design for their ablution platform with a consideration on the level of seat, drainage system and shelf. This design is the most recommended design for ablution platform with seat type. Model 3 is an ablution unit with lavatory, and it is almost similar to the design of lavatory in the home bathroom. Users need to bend to reach the faucet and raising their feet for the final ablution steps to the lavatory. Model 4 shows an ablution platform with a lowered lavatory level and raised faucets level. This design is recommended and users need to stand while performing their ablution process. The faucet level is quite high than common design to minimize the bend over while the ablution process. The barrier is designed as low as knee level to give comforts to users and they can manage to raise their feet to wash it at the faucet area. A shelf is also available for users' convenience, similar to model no 1 and 2.

Table 1 Conventional Ablution Design by Mokhtar (2005)

	Model 1	**Model 2**	**Model 3**	**Model 4**
Ablution Design	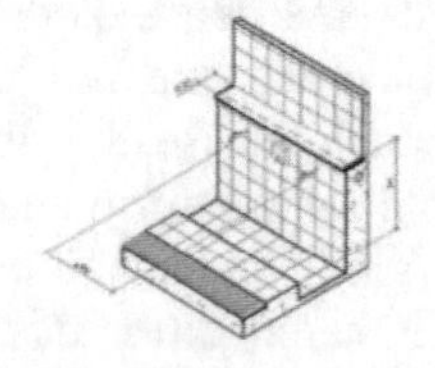	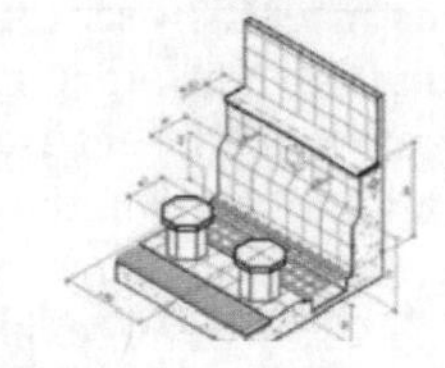	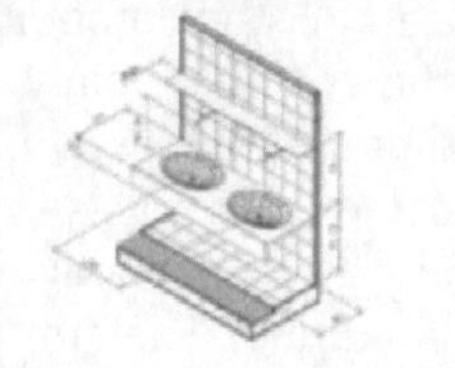	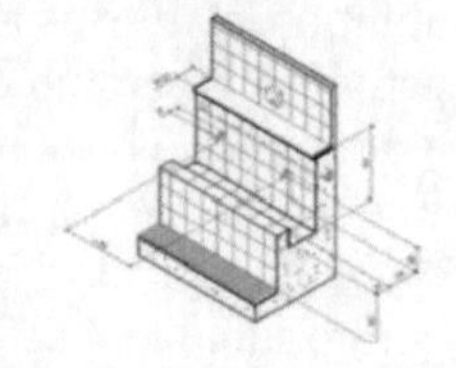
Features	Seats, shelf	shelf	shelf	shelf

Nashirudin (2008) proposed design guideline for ablution area in 3 kinds of aspects; 1) relation of anthropometric with the location, 2) area segregation, and 3) ablution area. The 1st aspect is anthropometric with the location, which also has strong relation with the ergonomic study. Therefore, each of stages in ablution process needs to be considered when designing the ablution area to answer the needs and demands of the users. The 2nd aspect is the location, which emphasized on the needs to separate ablution area for male and female, toilet, and to consider the 'Qiblah'. Area segregation focused on 3 areas; wet, dry/wet and dry to ensure the cleanliness and safety.

The 3rdaspect is the selection of the location of ablution area, which focuses on designing the distance and allocation of facilities in the ablution area; a distance of seats and faucet, and water performance. This aspect also has strong relation with the ergonomic study. Nashirudin (2008) also suggests the ablution area should focusing on the water performance to avoid a problem such as water split that causes wet to the clothes of users. According to Che Haliza binti Che Roza (2005), users do not alert on the excess of water usage during ablution process because of their low awareness on this issue. Zaini Ujang (2000) proposed the automatic faucet use in ablution platform by innovating the drainage system. In Islam, Muslim should use only 1 'mud' / 500 ml water for ablution to avoid excess usage of water.

Although various studies have been done previously, there are many mosques and 'surau' in Malaysia still do not apply as per suggested by the studies. This is maybe due to the lack of information, awareness and also the budget to renovate the ablution area. Improvement in the design of ablution area and spaces is needed by referring to the established guidelines. Furthermore, users need user-friendly facilities so that they get benefit whilst performing their ablution activity. These should include the needs of elderly and PwDs.

3. METHODOLOGY

This study employed qualitative methods of data collections. To determine the existing condition of physical elements of various ablution platform, an observation on physical elements of ablution platforms; ergonomically and user-friendly has been done in 15 mosques in Kuala Lumpur (KL) and Selangor, Malaysia; 1) Masjid Sultan Abdul Samad, 2) Masjid Sultan Salahudin Abdul Aziz Shah, 3) Masjid Al-Ikhlas, 4) Masjid Assalam, 5) Masjid Ara Damansara, 6) Masjid Al- Mukarramah, 7) Masjid Amaniah, 8) Masjid Wilayah, 9) Masjid At-taqwa, 10) Masjid Mujahideen, 11) Masjid Muhtadin, 12) Masjid Saidina Umar, 13) Masjid Nurul Islam, 14) Masjid Darul Ehsan,15) Masjid Bandar Utama.The mosques are under various categories hence the facilities and physical conditions of the ablution spaces reflect the size of the mosques. During the non-participant observation, photos of the ablution areas including its platform were documented for physical analysis.

4. RESULT, FINDINGS AND DISCUSSION

A total of 15 samples of ablution platforms from the 15 mosques have been analyzed as shown in Table 2. These include 4 mosques in Kuala Lumpur (KL) and 11 mosques in Selangor. Most of the mosques are under district mosque category. There is only one (1) institutional mosque which belongs to Kuala Lumpur International Airport (KLIA) known as Masjid Sultan Abdul Samad. Therefore, this mosque has different facilities in terms of sizes and numbers. The mosque is not only for the district population but also is also used as an international mosque for tourist who arrived and departs from/to KLIA. The results indicate that there are 2 kinds of ablution platform design; with seats and without seats. 11 mosques provide ablution platform with seats. Masjid Amaniah, Masjid Muhtadin, Masjid Nurul Islam and Masjid Darul Ehsan have ablution platform without a seat. The principal aspects of ergonomics and user-friendly were analyzed in all ablution platforms and spaces. The ergonomic study was done in terms of seat size, seat allocation, faucet allocation and height of the platform. Also, accessibility and a special platform for PwDs, floor condition and seat condition were observed. 10 out of 15 mosques have ablutions platforms with ergonomic physical elements. 10 mosques have their ablution areas with comfortable seats and 12 mosques have proper faucet allocation. A total of 11 mosques have non- slippery floors and seats. However, the ablution platforms in several mosques are provided with the lack of consideration towards the needs of PWDs. It was observed that the ablution areas in the 3 mosques are not accessible for PwDs and a total of 5 mosques have provided special ablution areas for PwDs.

Table 2: Physical features the ablution platforms and spaces of the selected mosques

		Ergonomic				**User-friendly**			
		Seats size	**Seats allocation**	**Faucet allocation**	**Platform height**	**Platform**	**Platform**	**Floor**	**Seats**
	Mosque	**Comfortable**	**appropriate**	**appropriate**	**appropriate**	**Accessible for PwD**	**Special for PwD**	**Not Slippery**	**Not Wet**
1	Mosque 1	√	√	√	√	√	√	√	√
2	Mosque 2	√	√	√	√			√	√
3	Mosque 3	√		√	√				√
4	Mosque 4	√	√	√	√	√	√	√	
5	Mosque 5	√	√	√			√	√	
6	Mosque 6			√	√		√	√	√
7	Mosque 7	√	√					√	√
8	Mosque 8	N/A (without a seat)			√			√	√
9	Mosque 9								
10	Mosque 10			√	√			√	
11	Mosque 11	√	√	√	√			√	√
12	Mosque 12	N/A (without a seat)		√	√				√
13	Mosque 13	√		√		√	√	√	√
14	Mosque 14	N/A (without a seat)		√					√
15	Mosque 15	N/A (without a seat)		√				√	√
	Score	10/15	8/15	12/15	9/15	3/15	5/15	11/15	11/15

Masjid Saidina Umar, which is located in KL has a spacious ablution area in which many ablution platforms with seats are provided (Figure 1a). In this spacious area, a bigger number of seats are installed, which can accommodate many users to perform their ablution activity at any one time. In addition, the arrangement of ablution seats is more ergonomic due to the spaciousness of ablution areas. Apart from ablution platforms, storage for keeping personal belongings also are provided in the ablution areas for user convenience and safety reasons. As mentioned by Ahmed Mokhtar (2005), the most efficient space planning need to have bigger space with different facilities to ensure the comfortability and safety of users. Meanwhile, Masjid Al-Muqarramah, which is located in Puchong, KL is provided with the ablution platform with seats (Figure 1b). However, it has been identified that the ablution seats are not ergonomic. The ablution seats have inappropriate dimensions and distance from the water faucet, as shown in Figure 1b.

Figure 1: Ablution spaces at a) Masjid Saidina Umar,b) Masjid Al-Muqarramah, and c) Masjid Al-Ikhlas

These conditions have contributed to the difficulty in ablution activities because the user needs to bend down and adjust his/her body posture to be able to perform the ablution. In addition, the distance between water faucets and seats are too narrow as well. In short, the ablution platforms; seats and faucet arrangement are found inappropriate thus users feel difficult in performing their ablution activity. This design is a low cost and uncomfortable for users to perform ablution because requires users to sturdily bend their back or knees. This kind of design is not user-friendly and yet dangerous because it can cause fall, especially to the PwDs.

5. CONCLUSION AND RECOMMENDATION

Many mosques in Kuala Lumpur and Selangor provide ablution platforms with seats for users' comfort and safety. Many mosques have consideration towards ergonomic ablution area including its ablution platform; arrangement of seats and faucet. However, there are several mosques that have unergonomic ablution areas, for example, the inappropriate height of water faucet that causes a difficulty to users to perform their ablution. In addition, most of the mosques do not provide special ablution platform or at least an accessible area for PwDs to perform their ablution activity. Most of the mosques were equipped with platform made from cement and tiles that are not slippery to ensure user's safety. The findings suggest that there is a need to adopt the design guideline proposed by Nashiruddin (2008). The design guideline has 3 aspects; the relation of anthropometric with the location, area segregation and ablution area, which also relates to the ergonomic and user-friendly concept. There is urgency for future research to examine the physical conditions of ablution platforms and spaces in other Malaysian mosques because it is vital for the mosques to provide good and safe facilities for worshippers to perform smooth ablution activity.

References

[1] A Rusmadiah Anwar, Shahriman Zainal Abidin,, Oskar Hasdinor Hassan. Function Means Analysis for Ablution Concept Solution. TOJET: The Turkish Online Journal of Educational Technology – September 2015, Special Issue for INTE 2015

[2] Mokhtar, A. (2005). Design guidelines for ablution spaces in mosques and islamic praying facilities. Journal of Architectural Engineering, American Society of Civil Engineers.

[3] Jasmani, Ismail and Awang, Arita Hanim and Ridzuan, Muhammad Ibrahim (2017) Interior design elements and facilities for elderly in the mosque: are we providing the needs? In: "The 3rd Putrajaya International Conference on Children, Women, Elderly and People with Disabilities 2017., 28-29 october 2017, Hotel Bangi Putrajaya, Selangor, Malaysia.

[4] Nashirudin M. A. S. & Jasmi K. A. (2008). Cadangan Penyediaan Tempat Wuduk Yang Efisien (1st ed.). Johor Bahru: UTM Press, pp. 124-140. ISBN: 978-983-52-0508-8.

[5] Ujang (2000) 'Water, Wealth dan Wisdom' di Universiti Kebangsaan Malaysia (UKM). kiktengah.uitm.edu.my/images/.../AQUA_MUTLAQ_Laporan.pdf

[6] A. Ghafar Ahmad. (1999). The Architectural Style of Mosques in Malaysia: From Vernacular to Modern Structures. Symposium on Mosque Architecture: The Historic and Urban Developments of Mosque Architecture, King Saud University, Riyadh, Saudi Arabia.

IOP Conf. Series: Materials Science and Engineering **401** (2018) 012021 doi:10.1088/1757-899X/401/1/012021

Light, daylighting and fluctuation of illuminance level in office buildings

Elina Mohd Husini[1], Raja Nur Syaheeza Raja Md Yazit[1], Fadli Arabi[1], Wan Norisma Wan Ismail[1] and Nor Haslina Jaafar[2]

[1] Architecture Department, Faculty of Engineering and Built Environment, Universiti Sains Islam Malaysia, Nilai, 71800, Negeri Sembilan, MALAYSIA
[2] Architecture Department, Faculty of Built and Environment, Universiti Kebangsaan, Bangi, MALAYSIA

elina@usim.edu.my

Abstract. Illuminance in spaces is important criteria in rating building performance. These also a strategy to develop some passive design criteria to meet the occupant visual comfort. The aim of this study is to investigate the fluctuation of the illuminance level and the impact of daylighting in an optimum performance zone. This study also investigates the effects of occupant density whether affect the illuminance level in two selected office buildings. Qualitative survey has been conducted on the occupant's perception toward visual comfort and obtaining the comparison on the glare effect by using IESVE computer simulation while visual test has been conducted involving selected workers. The maximum occupant density and physical conditions; window to wall ratio of 25%, 40% and 70% of typical office space were identified by using IESVE simulation. Findings from experiment of visual performance and daylighting simulation showed more than 50% occupancy in typical office room with 70% of WWR affected the extend of the daylight zone and the acceptable of illuminance level is not achieved. There high contract between 50% of the occupancy in a room is also changed the illuminance level of more than 26%. The findings contribute towards the improvement of an acceptable visual comfort ranges and design recommendations for optimum performance zone in a single office space in Malaysia.

1. Introduction

When most of the buildings were found to over illuminate or lack of illumination, it shows the designer should identify the significant parameter in lighting design during schematic layout till occupancy. This consideration and awareness can meet the acceptable illuminance level for occupants during performing the task less work for retrofitting such as de-lamping artificial lighting. Therefore, complaint by occupants on lighting distraction and too bright the illuminance level shows fluctuation of illuminance level occurs frequently in indoor spaces. In addition, these may reduce visual comfort and performance. Furthermore, the fluctuation of daylight illuminance level which affects the acceptable indoor daylighting level is not highlighted in any visual comfort issue where it significantly changes the quantity of daylight levels. This unpredictable change in illuminance level happens more severely when the room is fully occupied, and it is also reported that dimming control systems may be affected due to this condition. Clarification is needed on whether lighting guidelines have over- or underestimate the preferred amount of light. Factors that influence illuminance changes in offices are not identified during

Content from this work may be used under the terms of the Creative Commons Attribution 3.0 licence. Any further distribution of this work must maintain attribution to the author(s) and the title of the work, journal citation and DOI.
Published under licence by IOP Publishing Ltd

the measurement of quantity of illuminance level. The occupants are not aware of the illuminance change, and this illuminance variation may contribute to visual discomfort and visual annoyance. (Lee et al., 2012). This may also influence by physical condition as such as window size and people's preferences and satisfaction. Husini (2011), described the preferred illuminance level was related to occupant's satisfaction in daylighting condition. Meanwhile the current recommended settings of lighting have not yet reached the needed levels for occupants and optimum performance zone. A range of illuminance levels need to be furthered since it was detected unstable due to fluctuation of illuminance level. Until today there is no annual metric exists to access visual satisfaction in daylight environment and less work applying this metrics in daylight interiors (Rubiño et al., 1994; Jakubiec and Reinhart, 2012). In this paper, an investigation of fluctuation of the illuminance level and optimum occupancy in daylight zone is presented followed by a description on the fluctuation effect in minimum window-to-wall ratio. In this discussion section, only the range of the acceptable with the optimum window-to- wall ratio which meets the acceptable illuminance level is explored within the context of visual comfort methodologies. A review of the principle of light to human and the exploration of the ancient science of light from the previous Islamic scholars is explored in this study.

2. Principal of light, optic and visual performance

'Light' refers to the wave length that emphasis on the electromagnetic spectrum, which includes visible light, as well as light with wavelengths that we cannot see such as: radio waves, microwaves, infrared, ultraviolet, X-rays, and gamma rays. Light is the key ingredient which highlights the application in physics, environmental sciences pharmacology, engineering and industrialized design. The environmental implementation on light, daylighting and visual performance are strongly connected concept. These explorations on the overall atmosphere of the inner environment of the building is interpreted in a various transformation of architecture that started from early discovery of light. In the history of light, Ibn al-Haythm who is widely known as the Arab Scientist had demonstrated the experiment of light and he concluded that vision would only take place when a light ray which comes from a luminous source or was reflected from such a source before it entered the eye. His discoveries are truly extraordinary when he experimented to prove that we see because light from objects travels in a straight line into our eyes.

Further to the study on light, the illuminance is also an important part. Based on Ziai (2013) and Aminrazari (2014), the Shihab al-Din Yahya al-Suhrawardi, in his Kitab hikmat al-ishraq explained that Ishraq is commonly used to refer to the internal illumination and Illuminationism was exploring the ancient "science of lights" ('ilm al-anwar). It is described that the light and luminous in respect of degrees of intensity. It also provides luminosity, give shape to the forms and these make the entity 'visible and known. The light may cause the perceptual performance and visual comfort by occupants which causes the visual where there is a complex perception through human's senses. (Araji, 2008). This can adapt the wide range of brightness. Elzeyadi (2002) found that visual comfortable environment depends on vision perception, space configuration and corresponding to different activities. The eye receives light and sends signals to be processed by the brain through optical components (Araji, 2008). The role of lighting designers and architects necessitates balancing lighting conditions for optimal visual comfort in daylighting design. Several standards and guidelines in Malaysia are only considered when the electric lighting is used. Lighting quality as outlined in Malaysian Standard 1525:2014 is referring to this code of lighting practice to provide sufficient indoor lighting is at the range of 300 lux.

2.1. Illuminance, absorption, fluctuation and occupancy.

The light we see affects our perceived ideas of brightness. This light or amount of light reaching the eyes of the observer are considered luminance (IESNA Handbook, 2000). In this study, the illuminance level is focused on working plane illuminance and its impact on visual performance. The connection of the light and human occurred when there was light can be detected by the reflections of skin. The light that is not reflected at the surface, but penetrates the material body, will be scattered and selectively absorbed at wavelengths which provide the characteristic object color through body reflection. The incident of light (95%) entering the skin is mainly absorbed and scatters within two skin layers (body reflectance). (Storring et al., (2001). The theory of absorption by Lambert's Law explained that the basic principal of how absorption affects the light through some wavelengths while transmitting others. While the later theory based on Beer's Law elaborates on how the intensity of light decreases exponentially with depth in the material. Beer's Law states that a similar amount of absorbing material may also absorb an equal fraction of light. In this study, the concept of absorption is a focus where, in the transmission of light, an object can absorb part of the light and when this material absorbs some wavelength, it may transmit the light back by converting it into heat. Lighting represents almost 20% of global electricity consumption. The belief on light and work performance has also been studied by Hughes and Mc Nelis, (1978) and their studies have proved that illuminance affects task performance. Fluctuations in illuminance are identified through the impact from external and internal illumination. These fluctuations in illuminance will enhance the impact of indoor visibility and lighting systems. Light fluctuation affected the occupants when the room had a normal ambience of light, but the actual illuminance fluctuated over time and the desired illuminance could not be maintained as required. Fluctuation is known as brightness fluctuation, in which both the increasing light and decreasing light may determine different perceptions of lighting change. When they tested reading-based tasks using field measurements of fluctuation ranges or illuminance variation, their research showed a change in visual annoyance and visual comfort, resulting in a negative mood and lowered performance in the office. The factors that affect the fluctuation were mainly from the lighting from outdoor, sky condition and the window design. Boubekri, (1991) investigated the impact of window size and sun patches, finding that there was no significance from the latter to occupants' emotions, but that sunlight penetration through windows affects feelings. The consideration on guidelines and recommendation for daylighting and visual comfort is based on Universal Building Bye Law (UBBL) requires minimum Window to Floor Ratio (WFR) of 10 % for designing a space in a building. However, UBBL does not further outline any requirement for daylighting. (Yakubu, 2015). Another focus in the effect of absorption of light is the human occupancy in the space. Occupant density is defined as an ideally calculated in an office when the room is occupied. The indoor illuminance has a relationship with occupancy patterns when there are changes of illuminance level due to the crowding effect; this means that more occupants in a room will result in different energy consumption characteristics in different scales of office buildings. In short, a higher level of energy consumption occurs when the number of occupant's changes. The study from Elina (2016), concluded that when the number of occupants increases, the coefficient of variability decreases, and the lighting schedule was more regular.

3. Methodology

3.1. Experiment for visual test and computer simulation.

There are two phases in this study: (i) Experiment on visual test performance, and (ii) computer simulation. In the first phase is to analyze the experiment based on visual test on selected buildings; Malaysian-Japan International Institute of Technology, Universiti Teknologi Malaysia (MJIIT), Menara Tima and Midvalley office building. The final phase is conducted by using computer software; Integrated Environment Solution Virtual Environment (IESVE) software. A triangulation of the results concludes the research.

3.2. Experiment 1; Visual Test.

The purpose of conducting the visual test is to analyze the subjects' responses to capture the acceptable illuminance level in a space. The results were based on highest score and the lowest score which described in mean score. The test was referred to Roufs and Boschman (1997) who applied a visual test to determine text quality metrics for visual display units. The setting of the room was decided according

IOP Conf. Series: Materials Science and Engineering **401** (2018) 012021 doi:10.1088/1757-899X/401/1/012021

to the need for a visual quality aspect in a room which provided comfort for a reading task. The different occupancy patterns are according to the office ergonometric. This is to analyze visual comfort as a product of number of occupants and daylight distribution from the window and identify the ranges of preferred illuminance level and performance level achievement. A similar technique was applied by Wang, (2009) to capture sunlight patch in daylight zone.

3.3. Computer simulation; IESVE software

In this research, computer simulation was used to investigate the fluctuation in daylight illuminance with varying occupancy patterns. The computer simulation experiment in this study uses the IESVE software. Through the analysis of computer simulation, various designs such as room geometries, window-to-wall ratios, occupant density, and distance to window are considered as design variables. This will be a base setup for computer simulation using Integrated Environmental Solution Virtual Environment (IESVE) with different configuration of rooms such as linear room and deep room with varies of occupancy from 25%, 50% to 70% in window-to-wall-ratio of 70% (Figure 1). 70% of window –to-wall-ratio is considered as maximum opening which will provide stable daylit in indoor space.

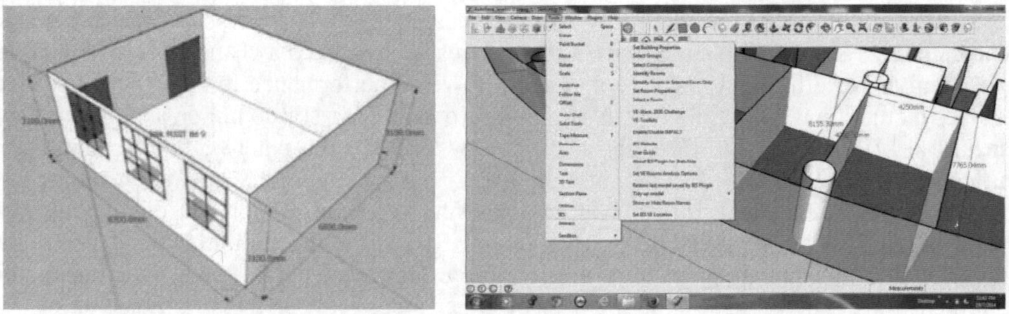

Figure 1. Simulation using IESVE Software in Selected Office Space

4. Results and discussion

4.1 Visual Test Result.

Table 1. Comparison of scores and speed and ranges of illuminance level

	Range of Acceptable illuminance level (in daylighting)	**Percentage Score (%)**	**The Fastest Speed (Time)**
TIMA Office	300-350 Lux	25% (2min – 3 min)	1min 35 sec – 1 min 45 sec
MIDVALLEY Office	401-499 Lux	25% (1 min 46 sec- 2 min)	1min 46 sec- 2min
MJIIT Office	401-599 Lux	50% (2 min 01 sec – 2 min 15 sec	1 min 16 sec-1 min 30 sec

The comparison of score and speed in visual test among three office buildings are shown in Table 1. It shows that 50% of occupants in the MJIIT Office Building were in the top group, with a speed between 2:01 and 2:15. This is followed by TIMA Office Building, which placed 25% of the subjects in the top group for scores in the visual test, which was conducted in the illuminance range of 300-350 lux. Midvalley office resulted in 25% of the subjects scoring in the lowest percentage, with illuminance levels of 401-499 lux. It is therefore shown that there was a correlation between visual performance and acceptable ranges of illuminance level

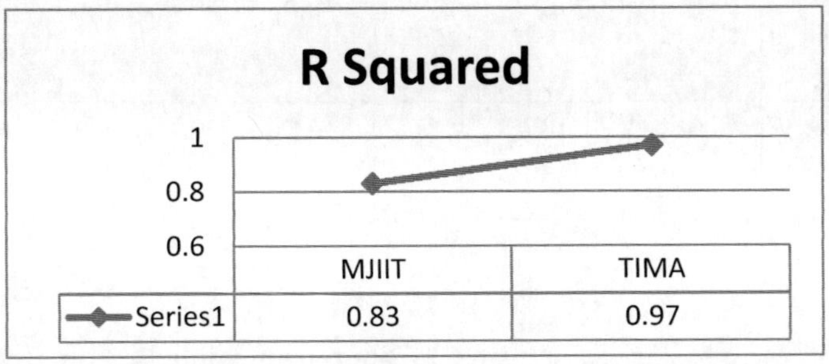

Figure 2. Percentage of Occupant and Scores of MJIIT Office Building

Figure 2 shows how the fluctuation of illuminance was significant, with a strong result of $R^2 = 0.83$ for Malaysian Japan International Institute Technology (MJIIT) and $R^2 = 0.97$ for Tun Ismail Mohamad Ali Office Building (TIMA)-. The illuminance level was reduced when occupancy was increased. When the maximum number of subjects occupied the daylight zone, the percentage of fluctuation - increased and these fluctuations happened.

4.2 Simulation Result

The simulation using IESVE software shows the daylight zone with minimum distance from window. The room was inserted with maximum number of occupant to identify the minimum distance to achieve acceptable illuminance level and which zone provides optimal task performance even there is fluctuation of illuminance. Figure 3 shows the 50% of the occupancy in a room changed the illuminance level of more than 26%. It means when this empty room showed 300 lux at the distance of 3075mm but when it was occupied by 50% of the room, the illuminance level fluctuated to 222 lux. The distance of daylight zone was also reduced from 4025mm to 2750mm when 50% to 75% occupancy in the room. This shows that 25% of the occupancy was considered the maximum occupancy in the room where the daylight zone can maintain the acceptable illuminance level of 300 lux. (-Refer to Table 2).

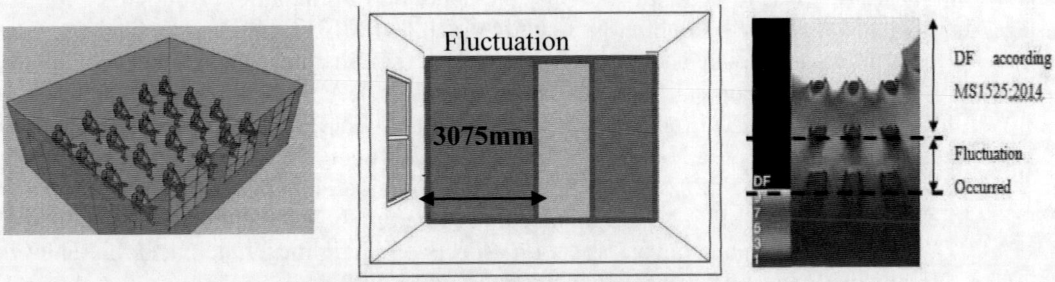

Figure 3. Daylight Zone and Distance

Table 2. Distance within daylight zone and acceptable illuminance range

Percentage of Occupancy	Daylight zone within acceptable illuminance range (distance)	Average illuminance level (lux)
0% occupancy	2.75 – 4.25 meters	406
25% occupancy	2075- 3075 mm	313
50% occupancy	2075 – 3000 mm	246.5
75% occupancy	2075– 3000 mm	125.6

Simulated the distance of daylight zone which was 3075mm from window with 70% window-to wall ratio was reduced to 2075mm distance from window in maximum percentage of occupancy. (-Refer Figure 3). The reduction will give the impact towards lighting decision and energy consumption. Besides that, the visual discomfort can be reduced.

5. Conclusion

This work concludes that the designer and architect need to organize the occupancy density by reducing or increasing the occupancy numbers from 25% to 50% in a room which can reduce the fluctuation by about less than 26% with the minimum daylight factor of 2% (250 lux). The fluctuation in daylighting was significantly detected from the experiment. It was also shown that the quantity of illuminance level, fluctuation and occupancy affected visual quality. The change pattern of average illuminance level at daylight zone was gradually decreased at certain occupancy.

Acknowledgment

This research is supported by the research Fundamental Research Grant Scheme (FRGS) 1/2015/TK07/USIM/02/1

References

[1] Aminrazari,M.A.R.(2014) Suhrawardi and the School Of Illumination, Routledge.
[2] Abu Sadin, M. F. M, Nik Ibrahim, N. L. and Sopian, K (2013) Ed in Azami Zaharim, Kamaruzzaman Sopian The Performance of External Shading Devices and Daylighting Rule of Thumb for a Tropical Climate Latest Trends in Renewable Energy and Environmental Informatics (res '13), (eninf '13) Volume 8 of Energy, environmental and structural engineering series pp 131-134 North Atlantic University Union; Kuala Lumpur, Malaysia, April 2-4, 2013 ISSN: 2227-4359 and ISBN: 978-1-61804-175-3
[3] Alrubaih, M. S., Zain, M. F. M., Alghoul, M. A., Ibrahim, N. L. N., Shameri, M. A. and Elayeb, O. (2013) Research and development on aspects of daylighting fundamentals Renewable and Sustainable Energy Reviews, 21(494-505)

[4] Arabi, F., Husini, E. M., and Kandar, M. Z. (2012).Effect of Natural Light Glare on Passive Design in Malaysian Government Office Building.American Transactions on Engineering & Applied Sciences.
[5] Araji, M. T. (2008). Balancing Human Visual Comfort and Psychological Wellbeing in Private Offices.ProQuest.Malaysia
[6] Aries, M. B., Veitch, J. A., and Newsham, G. R. (2010). Windows, View, and Office Characteristics Predict Physical and Psychological Discomfort. Journal of Environmental Psychology, 30(4), 533-541.
[7] Boubekri, M., Hull, R. B., and Boyer, L. L. (1991).Impact of Window Size and Sunlight Penetration on Office Workers' Mood and Satisfaction A Novel Way of Assessing Sunlight.Environment and Behavior, 23(4), 474-493.
[8] Elina Mohd Husini. 2016. Occupant Density And Daylight Illuminance Level Fluctuation For Office Building In Malaysia (Doctoral Dissertation).
[9] Elzeyadi, I. (2002). Designing for Indoor Comfort: A Systemic Model for Assessing Occupant Comfort in Sustainable Office Buildings. In Proceedings Of The Solar Conference (pp. 485-498). American Solar Energy Society: American Institute Of Architects.
[10] Hughes, P., & McNelis, J. (1978).Lighting. Productivity, and the Work Environment, Lighting Design+ Application, 8(12), 32-39.
[11] Husini, E., Arabi, F., & Kandar, M. (2011). On Post Occupancy Evaluation of the Preferred Luminous Environment and Occupants' Satisfaction for Office Buildings in Malaysia: A Survey.
[12] Jakubiec JA, Reinhart CF. The 'adaptive zone'–A concept for assessing discomfort glare throughout daylit spaces. Lighting Research and Technology. 2012 Jun 1;44(2):149-70.
[13] Lee, J. H., Yoon, Y., Baik, Y. K., & Kim, S. (2013). Analyses on human responses to illuminance variations for resident-friendly lighting environment in a small office. Indoor and Built Environment, 22(3), 535-550.
[14] Rubiño M, Cruz A, Garcia JA, Hita E. Discomfort glare indices: a comparative study. Applied optics. 1994 Dec 1;33(34):8001-8.
[15] Störring, M., Andersen, H. J., and Granum, E. (2001). Physics-based Modelling of Human Skin Colour under Mixed Illuminants. Robotics and Autonomous Systems, 35(3), 131-142.
[16] Yakubu Dodo. 2017.Visual Comfort (Doctoral Dissertation).
[17] Ziai, H. (2013). Shith al-Din Suhrawardi: founder of the Illuminationist. History of Islamic Philosophy, 434.

IOP Conf. Series: Materials Science and Engineering **401** (2018) 012022 doi:10.1088/1757-899X/401/1/012022

Thermal Comfort in Semi-Outdoor Studying Spaces: A case study of Universiti Sains Malaysia

M S OUAL, A A TABASSI and A S HASSAN

School of Housing, Building and Planning, Universiti Sains Malaysia, 11800 Penang, Malaysia

E-mail: seddikoual@gmail.com

Abstract. The paper presents an assessment of thermal conditions within semi-outdoor spaces used for the activity of studying. It aims to observe the potential of this type of spaces for human activities as a replacement or complementary to air-conditioned indoor spaces in order to reduce the energy consumption. The field work research, consisting of meteorological measurements alongside a questionnaire survey, was conducted in three studying areas at Universiti Sains Malaysia. The results demonstrate that under certain conditions, thermal comfort is achievable in studying semi-outdoor spaces. Further, the significance of objective and subjective factors in shaping users' thermal comfort perception is discussed.

1. Introduction

Building sector is witnessing a mechanization increase and rising of energy consumption. In tropical countries such as Malaysia, air-conditioning system becomes the primary consumption for occupants' thermal comfort. This problem is largely linked to the current approach of comfort that is based on the heat balance model. As a result of adopting this model, thermal comfort is specified within a narrow range, which is usually difficult to achieve without air-conditioning. Consequently, it promotes unsustainable extensive use of these systems. Because of that, current thermal comfort approach is one of the most controversial subjects in the field of building science [1]. Within the heat balance model, comfort is provided to the passive recipient occupant of the space. Comfort is defined by six factors affecting the thermal balance between the human body and his environment. Four factors are environmental (air temperature, mean radiant temperature, air velocity and relative humidity) and two are personal (metabolic rate and clothing insulation) [2].

The alternative view of comfort is the adaptive approach. Its principle contends that in the case of discomfort, people will act to restore their comfort. The occupant is considered more than a passive recipient of the thermal conditions. He rather plays an active role to achieve his comfort [3]. In this approach, comfort depends on the abovementioned factors in addition to other personal ones that dictate the different adaptation processes. These processes are categorised as follows: physiological adaptation (acclimatisation); behavioural adaptation (clothing change, changes in the activity level or taking hot or cold drinks, changing of position or posture, or moving to another place, etc.); and psychological adaptation (experience, expectations, time of exposure, perceived control, naturalness, environmental stimulation). This adaptive approach allows a much wider thermal comfort range compared to the heat balance model. Such a distinction opens opportunities to take advantage of non-mechanical solutions to create comfortable environments. One of these opportunities is the reengagement with the outdoor and

Content from this work may be used under the terms of the Creative Commons Attribution 3.0 licence. Any further distribution of this work must maintain attribution to the author(s) and the title of the work, journal citation and DOI.
Published under licence by IOP Publishing Ltd

IOP Conf. Series: Materials Science and Engineering **401** (2018) 012022 doi:10.1088/1757-899X/401/1/012022

softening the barriers between indoor and outdoor [4, 5]. In this connection, semi-outdoor spaces could play such a role. They are defined as spaces open to the outdoor while still not open to the sky [6].

The literature shows that the semi-outdoor spaces have potential for energy saving. They could be utilized as buffer zones and transitional spaces that offer time for thermal adjustment. These spaces could also provide a more comfortable conditions in contrast to the outdoors allowing interactions and engagement in certain activities. The energy demand, therefore, will be reduced as people spend less time indoors. The literature has mainly investigated thermal comfort in semi-outdoors being used for recreational activities, temporary activities or transitional spaces, including lobbies, balconies, pavilions, stations, etc. Nevertheless, few studies have shown that under the adaptive comfort approach, thermal comfort is achievable in semi-outdoor spaces—particularly in the tropical region— if some elements like shading and good air movement are taken into consideration [7, 8].

Based on these previous findings, the current paper aims to further assess thermal comfort in semi-outdoors used for a long-term activity usually held indoors, then examine the different factors affecting it. By playing the role of an alternative or complimentary space to a mechanically controlled indoor environment, a conclusion of whether these spaces may reduce energy consumption could be drawn. The activity of studying is selected as a case study. Besides the possibility of energy savings, using semi-outdoors settings for studying can be very beneficial to the learning process [9].

2. Materials and method

This study adopts a quantitative approach based on micro-climate field measurements alongside a transverse questionnaire.

2.1. *Study location*

The field work is conducted at Universiti Sains Malaysia, in Penang, Malaysia. Within USM campus—based on attendance rate and physical characteristics—two semi-outdoor spaces used for studying have been selected. These are the semi-outdoor reading area of the library (Code: L2) and the dining hall of RESTU and SAUJANA hostels (Code: L3). L2 is an open, naturally ventilated space, whereas L3 is an open space with mechanically aided ventilation using ceiling fans. Additionally, a third space, the 24-hour reading room, an air-conditioned indoor space, is selected as a reference for comparison (figure 1).

2.2. *Collected data*

A number of thermal indices can be employed to evaluate thermal comfort. For outdoor and semi-outdoor studies, the physiological equivalent temperature (*PET*) is considered to be the most suitable. In order to calculate *PET*, RayMan software version 1.2 is used, which requires the input of six parameters: air temperature (T_a); humidity (*RH*); air velocity (V_a); mean radiant temperature (T_{mrt}); activity level (*Met*); clothing insulation (I_{cl}). The first three parameters are recorded automatically on-site using a microclimatic station, which is composed of two data loggers connected to the different sensors, all mounted on a 1 m height tripod (figure 1). T_{mrt} is calculated based on measurements of globe temperature (T_g). In order to measure T_g, a globe thermometer have been created using a (Pt100) temperature sensor, positioned in the centre of a 40mm acrylic ping pong ball painted flat grey (RAL7001) [10]. *Met* is set to 58.2 W/m2 for reading and writing activity. I_{cl} is calculated for each respondent based on his/her clothing items.

The questionnaire consists of three sections. The first collects respondents' demographic information, the second concerns subjects' perception of comfort level and the third section examines the influence of different objective and personal factors on comfort.

Figure 1. Locations and instruments. (L1) 24-hour reading room. (L2) Semi-outdoor reading area of the library. (L3) Dining hall of RESTU and SAUJANA hostels. (a) air velocity . (b) air temperature and humidity . (c) globe temperature. (d & e) data logger.

2.3. *Procedure*

The study was conducted at the end of a February month over a period of 5 days, with 15 hours surveyed each day (8 a.m. to 23 p.m.), according to the following procedure: While the measurements are collected automatically with a sampling rate of 5 minutes, the arrival time to the surveyed area of every person is noted, then after 15 minutes—allowing adjustment to the space conditions [11]—the subject is approached for participation. Following these steps, every user of the space is approached, except people staying less than 15 minutes and those using the space for activities other than studying. At the end, 348 questionnaires have been collected to be analysed and compared to the measured parameters.

3. Results and discussion

3.1. *Thermal comfort state*

The highest thermal comfort votes are observed among the occupants of semi-outdoor reading area (L2), followed by the 24-hour reading room (L1) then the dining hall (L3) (figure 2). 82.6%, 79.1% and 57.1% of occupants' votes are within "neutral" to "very comfortable" range, corresponding to (M=0.70, SD=1.29), (M=0.63, SD=1.32) and (M=-0.32, SD=1.12) in L2, L1 and L3 respectively. Furthermore, it is observed that in the case of L2, comfort level is even better than in the air-conditioned indoor space. In what follows, we will explore the different factors that have led to these results.

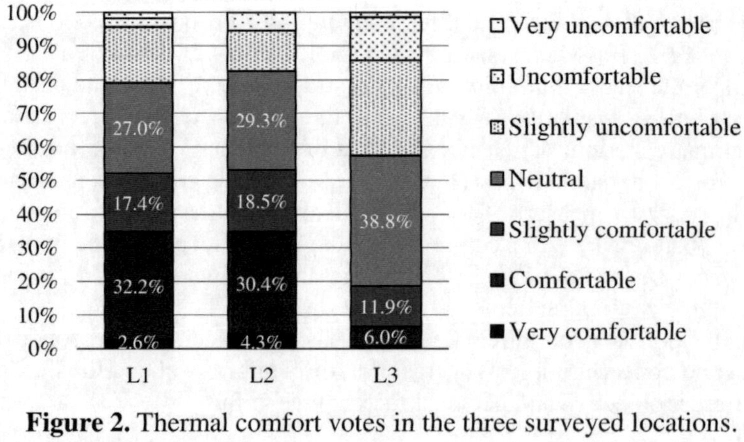

Figure 2. Thermal comfort votes in the three surveyed locations.

3.2. *Environmental factors*

3.2.1.Air temperature (T_a). The recordings show that the lowest air temperatures are measured in L1 (M=21.41 °C, SD=1.28), where the temperature has varied between 19.42 and 23.25 °C. The semi-outdoor L3 (M=30.55 °C, SD=1.81) occupies the second place with a wider variation range (26.92–

33.88 °C). Lastly, L2 (M=31.01 °C, SD=2.19) with temperatures varying from 26.42 °C to 34.16 °C (figure 3). These results do not exactly match the respondents' temperature perceptions, where on a scale from "Cold" (-3) to "Hot" (+3), L1 is perceived to have the lowest temperature (M=-1.10, SD=1.09), followed by L2 (M=0.17, SD=1.23) then L3 (M=0.96, SD=1.29). Moreover, thermal satisfaction that corresponds to the thermal sensation votes (TSV) within the range "slightly cool" (-1) to "slightly warm" (+1), is higher in the semi-outdoors L2 (72.8%) and L3 (65.7%) compared to the indoor L1 (62.6%). By juxtaposing these results with thermal comfort votes, it is clear that the latter correspond better with the perceived temperature rather than the measured. This is confirmed by the correlation tests (table 1).

3.2.2. Relative humidity (RH). The recorded relative humidity is approximately constant in L1 (45–50%). However, in L2 and L3, humidity varies following the opposite direction of air temperature, with (36–63%) in L2 and (42–73%) in L3. Furthermore, the votes regarding humidity, on the seven-point scale ranging from "Much too dry" (-3) to "Much too humid" (+3), are mostly within the "just right" category. That is 61.7% of L1 users, 60.9% in L2 and 41.4% in L3. The rest of the votes are concentrated on the dry side of the scale i.e. (M=-0.23, SD=0.74) in L1, (M=-0.32, SD=0.73) in L2 and (M=-0.26, SD=1.16) in L3.

A correlation of data shows that a significant relationship exists between *RH* and thermal comfort votes in the studied semi-outdoors. High *RH* levels are associated with low comfort levels. In contrast, humidity perception is not significantly related to comfort level (table 1). According to Metje, Sterling and Baker [12], this insignificance is due to peoples' inability to realise the effect of humidity unless it is very high or very low.

3.2.3.Air velocity (V_a). The highest levels are registered in L2 (M=0.82 m/s, SD=0.51) followed by L3 (M=0.44 m/s, SD=0.13) then L1 with low and stable levels (M=0.11 m/s, SD=0.04). The maximum air velocity is recorded in the semi-outdoor L2, which has the value of 1.86 m/s (figure 3). These results are in accordance with respondents' evaluations, where air velocity is perceived to be on the breezy side of the scale in L2 (M=0.40, SD=0.85). On the other hand, the conditions tend toward the still side in L1 (M=-0.31, SD=0.82) and L3 (M=-0.17, SD=0.99). In addition, a preference of "more" air movement in the semi-outdoor spaces is observed, especially in L3 where 64.2% of users want "more" air-movement compared to 57.6% in L2. In L1, the majority (50.9%) are content with the situation and want "no change".

The correlations (table 1) demonstrate that both measured and perceived air-velocity are significantly related to thermal comfort votes. Higher air velocity is associated with better thermal comfort levels.

3.2.4.Radiant temperature (T_{mrt}). The mean radiant temperature does not significantly differ from air temperature. In the semi-outdoor spaces, the difference between the averages is 0.38 °C. The most noticeable difference is during the afternoon period where T_{mrt} is a bit higher. In the indoor space T_{mrt} is higher than T_a throughout the day, with an averaged difference of 0.89 °C. The seemingly small difference between T_a and T_{mrt} could be explained by the absence of direct sun exposure, which minimises the radiant temperature exchange. Regarding solar radiation perception, on a seven-point scale ranging from "very weak" (-3) to "Very strong" (+3), the votes tend toward the "strong" side of the scale in L3 (M=1.08, SD=1.31) and L2 (M=0.65, SD=1.19). In the indoor space L1, however, the votes tend toward the "weak" side (M=-0.14, SD=1.43). Furthermore, regarding preferences—except for the "no change" category that is chosen by 50%, 46.2% and 36% of L1, L2 and L3 occupants—the other categories ("more" and "less" sun) are balanced to a certain degree. This is most probably due to users' different exposures (i.e. users on the perimeter are more exposed than those in the centre of the space).

In relation to thermal comfort votes, it is found that the measured *Tmrt* does not significantly correlate (table 1). Contrariwise, users' perception of solar radiation is significantly related to comfort level.

3.2.5.Physiological equivalent temperature (PET). In the semi-outdoor spaces, *PET* varies between 24 and 34 °C (figure 3), with a neutral value of 27.4 °C (figure 4). Neutral *PET* is calculated for thermal sensation votes between (-1) and (+1) based on the method described in Lin and Matzarakis [13]. *PET* variation is generally close to T_a or a bit higher. However, during the morning period in L2—as a consequence of high air velocity—*PET* is much lower than T_a, where the difference reached a maximum of 2.59 °C.

The calculated *PET* is compared to thermal comfort ranges of Western/Middle Europe (18–23 °C), and sub-tropical Taiwan (26–30 °C) [13]. In the first comparison, despite *PET* laying outside the comfort range, the current study shows that comfort level is high in the thermally better space L2. It clarifies that occupants of the studied spaces are more tolerant to higher temperatures compared to the European study. This indicates that adaptation is taking place. The second comparison shows that despite the very close neutral *PET* values between the present study (27.4 °C) and Lin and Matzarakis' research (27.2 °C), within the reference comfort range (26–30 °C), thermal acceptability in L2 is only 53.8% as opposed to 88% in the case of Lin and Matzarakis [13]. Acceptability in this case is taken as the "no change" category in temperature preference votes [7]. Even when considering acceptability as satisfaction (-1<TSV<+1), it is 82.1%, which is still lower than Lin and Matzarakis' (2008) results. This indicates that the respondents are less tolerant than what is estimated by the sub-tropical comfort range. Such an outcome could be linked to the investigated activity (i.e. studying), which requires more favourable conditions compared to outdoor recreational activities of the referenced study.

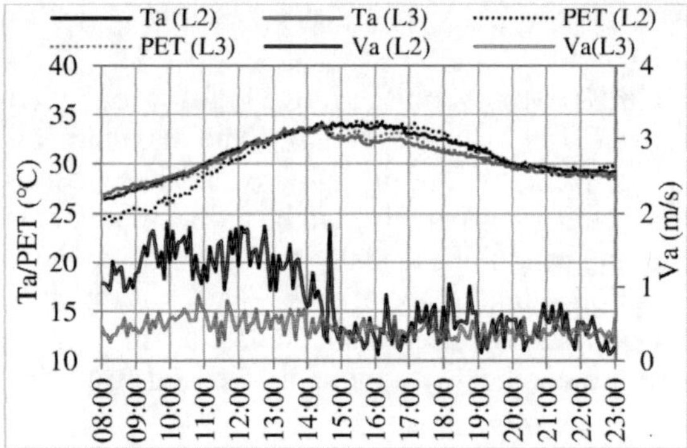

Figure 3. The physiological equivalent temperature, air temperature and air velocity variations in the studied spaces.

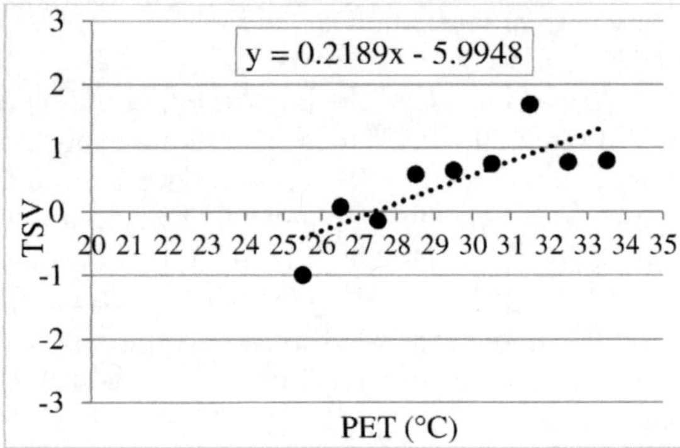

Figure 4. Thermal sensation votes and neutral *PET*.

3.3. *Personal factors*

3.3.1. Gender. An independent-samples t-test is conducted to compare thermal comfort votes between male and female participants. The result shows no significant difference in the scores for males and females (table 2). Gender does not have an effect on thermal comfort votes.

3.3.2. Ethnicity (Locals vs. Foreigners). Comparing thermal comfort votes between local and foreign respondents using t-test shows no statistically significant difference (table 2). Another test of thermal preference, however, shows a significant difference in the scores of locals (M=-0.54, SD=1.43) and foreigners (M=-1.73, SD=1.19); t (224)=2.71, p=0.007. This suggests that ethnicity has an effect on thermal preference, where foreigners prefer lower temperatures compared to locals. The findings are in line with attendance numbers in the adjacent spaces L1 and L2. During the two day investigation in L2 (warm semi-outdoor), it has been used by 9 foreigners and 83 locals. On the other hand, L1 (cold indoor) has been used by 39 foreigners and 76 local students during one day.

3.3.3. Cooling method preference. In response to cooling method preference, 13.7% of semi-outdoor respondents has chosen natural ventilation, 62.4% natural ventilation with fans and 23.9% air-conditioner. A one-way ANOVA is carried to evaluate the differences in thermal comfort votes based on participants' cooling method preference. The results illustrate a significant difference between groups; F (2,223)=6.636, p=0.002. A Tukey post-hoc test has revealed that thermal comfort votes are higher for participants who prefer natural ventilation (M=0.48, SD=1.31, p=0.004) and natural ventilation with fans (M=0.21, SD=1.26, p=0.006) compared to those who prefer air-conditioner (M=-0.43, SD=1.22). There is no statistically significant difference between those who prefer natural ventilation and those preferring natural ventilation with fans (p=0.507). These results point to the changes in people's preferences and thermal comfort requirements, where they are becoming less tolerant to natural conditions due to their habituation to the air-conditioner.

3.3.4. Home temperature. A correlation of the data reveals that the perceived home temperature and thermal comfort votes are not significantly related (table 1). However, a significant correlation is found between the perceived home temperature and temperature preference, r=0.17, N=226, p=0.01. High temperature at home is associated with high preferred temperature.

3.3.5. Temperature in the previous space. Using t-test, this study has found that temperature of the previous space does not have an impact on thermal comfort votes. Comfort votes of participants "who were in a hotter space" does not differ significantly from those "who were in a colder space" (table 2). The adjustment time (15 minutes) given to the participants before answering the questionnaire could explain this result.

3.3.6. Expectations. A significant correlation between thermal expectations and thermal comfort votes is found (table 1). The closer the actual thermal conditions to respondents' expectations are, the higher the thermal comfort votes.

3.3.7. Time of exposure. In response to the question about the time to be spent studying in the two semi-outdoor spaces, the answers have varied between 0.5 hour and 12 hours, with an average of 2 hours and 18 minutes. However, no significant correlation is found between thermal comfort votes and the time to be spent in the area (table 1).

3.3.8. Perceived control. Using t-test, the comparison of thermal comfort votes between students working in a group and those working individually shows a significant difference (table 2). Students working in a group, where leaving the space in case of discomfort is partially controlled by the group, have lower thermal comfort votes compared to those studying individually—they have control over the decision to leave the space. This suggests that having control leads to better comfort.

3.3.9. Naturalness and environmental stimulation. Among the multiple reasons for choosing the studying areas (temperature, light, quiet, etc.), environmental beauty has been chosen by 22.6% of the respondents in the semi-outdoors spaces. A t-test shows a significant difference in the scores of thermal comfort votes between the respondents who consider the beauty of the environment as an influencing element in their space choice and those who don't (table 2). Participants who have chosen this element feel more thermally comfortable.

3.3.10. Behavioural adaptation actions. Among the different actions that could be taken to improve thermal comfort within the studying area, the respondents in the semi-outdoor spaces have chosen consuming cool drinks (64.2%), changing the area (52.7%), changing seating place (27.4%), extra piece of clothing (19.5%), USB fan (8.8%), hot drinks (3.5%) and other solutions (1.8%). In contrast, in the indoor space, the most prominent action has been wearing extra piece of clothing (66.1%), followed by

IOP Conf. Series: Materials Science and Engineering **401** (2018) 012022 doi:10.1088/1757-899X/401/1/012022

changing the area (33%), hot drinks (23.5%), changing seating place (20%), cool drinks (18.3%), USB fan (4.3%) and other solutions (1.7%).

Further analysis of respondents' actual clothing level (I_{cl}) in relation to *PET* shows that in the semi-outdoor spaces, I_{cl} does not correlate significantly with *PET* (r=-0.08, n=205, p=0.24). In contrast, in the indoor space, a significant correlation between *PET* and clothing worn by participants is found (r=-0.19, n=115, p=0.038). Low temperatures are associated with more clothing. In semi-outdoors' case, the insignificance may be linked to the fewer possibilities to change clothing due to the high temperature levels as participants already wear the minimum they can. In contrast, lower temperatures in the indoor space allow occupants to wear extra cloths or remove them in case of discomfort.

Table 1. Thermal comfort correlation with environmental and personal factors.

Pearson correlation		T_a	Temperature perception	*RH*	Humidity perception	V_a	Air velocity perception
Thermal comfort votes	r	0.05	-0.32	-0.31	0.05	0.38	0.31
	p	0.50	<0.01*	<0.01*	0.48	<0.01*	<0.01*
	N	226	226	226	225	205	226

Pearson correlation		T_{mrt}	Sun radiation perception	Home temperature perception	Expectations	Time of exposure
Thermal comfort votes	r	-0.02	-0.33	0.09	0.28	0.07
	p	0.75	<0.01*	0.19	<0.01*	0.31
	N	226	226	226	225	217

Table 2. Independent-samples t-test results.

T-test (Thermal comfort votes)		N	Mean	Standard deviation	t	df	*p*
Gender	Male	56	0.04	1.36	-0.382	224	0.703
	Female	170	0.11	1.27			
Ethnicity	Local	215	0.07	1.28	-1.193	224	0.234
	Foreigner	11	0.55	1.44			
Previous space is hotter	Yes	82	0.24	1.38	1.410	223	0.160
	No	143	-0.01	1.22			
Studying environment	Individually	70	0.46	1.47	2.756	219	0.006*
	Group	151	-0.5	1.18			
Naturalness	Yes	51	0.63	1.28	3.440	224	0.001*
	No	175	-0.06	1.26			

4. Conclusion

Based on the current findings, it could be concluded that thermal comfort is achievable in semi-outdoor spaces used for the long-term activity of studying. Furthermore, under suitable conditions, thermal comfort level in these spaces can even surpass an air-conditioned indoor. The suitable conditions include both objective and subjective factors. Besides air velocity being the most notable environmental factor objectively affecting thermal comfort votes, this research has shown the subjective factors having the most significant impact. In this connection, occupants' perceptions of the different environmental conditions correlated better with comfort votes than the measurements did. Further, the different adaptation mechanisms are shown to be affecting thermal comfort votes' outcome or participants' thermal preference, including past experiences, expectations, perceived control and naturalness of the environment.

Acknowledgements
This work was supported by SKIM GERAN PENYELIDIKAN FUNDAMENTAL (FRGS), No. 203.PPBGN.6711515.

References

[1] Chappells H and Shove E 2005 Debating the future of comfort: environmental sustainability, energy consumption and the indoor environment *Building Research & Information* **33** 32-40
[2] O. Efeoma M, K. Ahadzie D, A. Ankrah N and Uduku O 2014 Assessing thermal comfort and energy efficiency in tropical African offices using the adaptive approach *Structural Survey* **32** 396-412
[3] de Dear R and Brager G S 2001 The adaptive model of thermal comfort and energy conservation in the built environment *Int J Biometeorol* **45** 100-8
[4] Wilhite H 2009 The conditioning of comfort *Building Research & Information* **37** 84-8
[5] Cooper G 2008 Escaping the house: comfort and the California garden *Building Research & Information* **36** 373-80
[6] Sinou M and Steemers K 2004 Urban semi-enclosed spaces as climate moderators. In: *The 21th Conference on Passive and Low Energy Architecture,* (Eindhoven, The Netherlands pp 1-5
[7] Hwang R L and Lin T P 2007 Thermal comfort requirements for occupants of semi-outdoor and outdoor environments in hot-humid regions *Architectural Science Review* **50** 357-64
[8] Ahmed K S 2003 Comfort in urban spaces: defining the boundaries of outdoor thermal comfort for the tropical urban environments *Energy and Buildings* **35** 103-10
[9] Hani N, Nasir M, Salim F and Yaman M 2014 The potential of outdoor space utilization for learning interaction. In: *UMRAN2014,* (Kuala Lumpur: International Islamic University Malaysia) pp 1-17
[10] Thorsson S, Lindberg F, Eliasson I and Holmer B 2007 Different methods for estimating the mean radiant temperature in an outdoor urban setting *Int J Climatol* **27** 1983-93
[11] Givoni B and Noguchi M 2004 Outdoor comfort responses of japanese persons. In: *The 21th Conference on Passive and Low Energy Architecture,* (Eindhoven, The Netherlands pp 19-22
[12] Metje N, Sterling M and Baker C J 2008 Pedestrian comfort using clothing values and body temperatures *Journal of Wind Engineering and Industrial Aerodynamics* **96** 412-35
[13] Lin T P and Matzarakis A 2008 Tourism climate and thermal comfort in Sun Moon Lake, Taiwan *Int J Biometeorol* **52** 281-90

IOP Conf. Series: Materials Science and Engineering **401** (2018) 012023 doi:10.1088/1757-899X/401/1/012023

Evaluating Public Expectation on Physical Factors of Community Center Designs in Kuala Lumpur, Malaysia

Nangkula Utaberta[1,2], Azmal Sabil[1,3] and Nastaran Jafari [1]

[1]Department of Architecture, Faculty of Design and Architecture, Universiti Putra Malaysia
[2]Halal Products Research Institute, Universiti Putra Malaysia
[3]Department of Architectural & Design Engineering, Faculty of Civil and Environmental Engineering, Universiti Tun Hussein Onn Malaysia

Abstract. There is a number of factors have to be considered in designing a public building or public spaces as it involves a wide range of users. In this paper, one of the important public spaces will be discussed, which is community center buildings. Community center as a public space must be seen and treated as a crucial part of the planning process for urban development. The Malaysian government is consistently trying to improve the quality of life with Malaysian's aspiration toward becoming a developed nation. The main objective of this paper is to identify the significant criteria for formulating a design framework for responsive community centers based on community expectation in the Malaysia. To achieve this objective, four community centers in four regions of Kuala Lumpur were used as the case studies in this paper. As one of the key elements in the building design, the physical factors of community centers chose as a main discussion thoroughly in this paper. This research was conducted in Quantitative research orientation, where the data was collected through questionnaire survey and analyzed by using SPSS. The outcome of this paper will be bridging the step toward the development of holistic framework and guideline in designing responsive community center.

Keyword: Expectation, physical, Community Center

1. Introduction

There are numbers of criteria in defining the physical characteristic of the building. It includesthe ventilation and lighting quality, size and the overall appearance of the buildings. The quality and value of architectural designs have been a matter of debate since the ancient Greeks organised a design competition in448 BC for the design of a war monument at the Acropolis(Glover, 2004). Nowadays, there are buildings with mixed usage such as: combining commercial, leisure, educational functions and unneeded buildings being converted to appropriate current requirement by changing their use. The method that acting is responded consists of: attention of common –day trend, opinions and political structures, the meaning of historical record, the tendency to present symbolic magnitudes and the vastness of common civil applicable to development (Tunstall, 2012). Although, the company of elements can be the design of one building type, it is important all of these elements work together correctly and expose the true messages.

The quality of the building is a major determinant of the success of the center, has straight relation to the achievement of the community center and a good social life is dependent upon a good building (Mess and King 1947). Juhari, Ali, & Khair, (2012) present arguments to emphasize that, the most public space does not take note of the environmental impact on the behavior of participants when, in reality, the environment will contribute to the number of participants visiting the place. In this way, it is said that the experience and background is a suit guideline for creating new success

Content from this work may be used under the terms of the Creative Commons Attribution 3.0 licence. Any further distribution of this work must maintain attribution to the author(s) and the title of the work, journal citation and DOI.
Published under licence by IOP Publishing Ltd

buildings in order to avoid the prior mistake. So it will be resulted via understanding the appropriate needs and the aim of each building for using the correct elements carefully. Finally, these can be sorted into the following wide series (Tunstall, G.2012).

2. Literature

A community center should be regarded as an essential amenity of normal community living in normal circumstances (Smith, M. K, 2002). As in Malaysia, there are various conceptual definition and function of community center. According to DBKL (2004), The design of the community center provides a hall, which can be used at any time for recreational activities or social gathering. These community centers do not meet with people's aspiration and choices which are being more diverse. Nurul et al (2014) declared that, during her survey observation, the researcher visited the recreational facilities like the community center around the study area for several times and different times. She found the few of them that are provided, but underutilized. This situation has arisen due to inadequate function and facility which are provided in the community center such as children's playground, soccer fields, sport facilities, library, kindergarten and other activities. Beside this, poor accessibility and inappropriate condition of building like ventilation. This is also acknowledged by claims highlighted in the National Urbanization Policy. Utaberta et al, 2010 acknowledged that, the community center of Kuala Lumpur is nothing more than a large empty space, used only for specific purposes like marriage ceremony or birthday ceremony and games like badminton, ping pong; it faces with the shortage of activity which held on at the community center.

There are many factors affecting the efficiency of community center. Whyte (2000) considered that, four key qualities play significant role in community centers as public space efficiency as: being accessible, individual being engaged in activities, the space being comfortable and has a proper vision, and it is a sociable space for people meeting in order to get together and enhance their social interaction.

Abidin et al (2005) presented arguments to emphasize that the life and activity that take place in space specify the usage and aim of making it or function. With this background, the main aspects that influence the use of public space are the form, location, physical and spatial structure as a physical factor of the community center. Smith et al (1997) additionally explained that professional design as physical properties that contribute to the quality of a community center, which play a more influential role in development and usability of successful community center. Jacobs (1961) in many years ago concluded that, the physical quality of public space can encourage people to join in a diversity of activities during the time of day and even at night. This, in turn, creates a safe and pleasurable environment, which functions, on the one hand, to reproduce existing social relations and facilitate community bonding and, on the other hand, to create the conditions to support local economic activity. As such, the economic potential of public space is entwined with and may even be dependent on social and environmental features (Grodach, 2009). In that sense, it concludes the importance of physical factor in determining the efficiency of the community centre.

3. Methodology

In this study quantitative methodology was chosen and applied to study the people's evaluation and perception toward using community center. The suitability of quantitative methodology via introducing it as scientific elicit of phenomena related using numerical data was declared by Muijs (2004) and Thomas (2003). Thomas (2003) expressed that; the result of quantitative study can be simply generalized over an enormous population.

A methodological framework is framed the variables and principles which are probed in a systematic manner. This framework contains the variables which must be assayed, the methods used to gather data and analyzing it. The principals being evaluated including the participant's expectation on the accessibility and the location selected for community centers in 4 regions of Kuala Lumpur. Questionnaire survey is used to evaluate the current situation of community centers. This method is suitable because the participants evaluate the current condition of these community centers on why they are not usable for publics.

IOP Conf. Series: Materials Science and Engineering **401** (2018) 012023 doi:10.1088/1757-899X/401/1/012023

The questionnaire distributed among 330 participants in four community centers located in Kuala Lumpur. It is Community Centers of Wangsa Maju-Maluri, Community Centers of Sentul-Menjalara, Community Centers of Damansara-Penchala and Community Centers of Bukit Jalil-Seputeh. These places are selected based on functional characteristics that present the current characteristics of Kuala Lumpur community centers as highlighted in DBKL report. The questionnaire was designed based on the following criteria:

- Size of community center
- Appearance
- Form of the building
- Good condition of ventilation

A five-point Likert-type scale was used. Then, the obtained data were accumulated and analyzed by SPSS software. As the main objective focuses on participant's evaluation, the collected data would be shown in numerical and statistical analysis.

4. Result and Analysis

Table 1 shows the mean analysis of the participant's expectation toward physical factor. According to the table in terms of physical factors,'Good condition of ventilation (mean= 4.51, sd= 0.65)' is received the highest expectancy among the other factors. The second position is allocated to 'Attractive appearance (mean=4.23, sd=0.77. Meanwhile 'Form of the building (mean=34.04, sd=0.88)' and 'Size of the community center (mean=4.04, sd=0.81) are collected the minimum expectancy among others.

Table 1: Descriptive Statistics for physical factor

Variables		**Mean**	**Stdn**
Size of community center	330	4.04	.81
Appearance	330	4.23	.77
Form of building	330	4.04	.88
Good condition of ventilation	330	4.51	.65

(1= Not Important, 2= Less Important, 3= Natural, 4= Important, 5= Very Important)
(Source: Field Survey, 2015)

Next figure has given more detail about people's expectancy of the Kuala Lumpur to the size of community center as a building design in below. As can be seen in the figure 1, for 48.8% (n=161) of participants, size of community center had an important position in their mind. For 30.3% (n=100) of people enough space in community center was very important and it was one of the priorities. While 0.6%(n=2) of participants didn't believe in importance of size in community center for increasing usability of it, 3.6% (n=12) of participants had less expectancy toward the importance of size as well. Also, some participants did not offer any idea by 16.7% (n= 35).

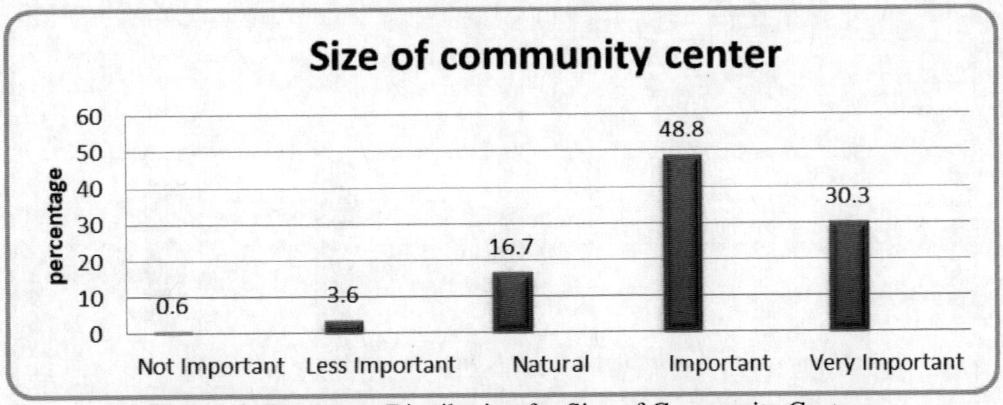

Figure 1. Frequency Distribution for Size of Community Center
(Source: Field Survey, 2015)

The next question and related figure (Figure 2) belong to appearance of the community center. For the majority of participants (45.5%, n=150) the appearance of the building had an important level. It is followed by 40.5% (n=134) of participants, which is the appearance of the community center were very important for them. Meanwhile, only one person (3%) didn't any expectation about the Attraction of elevation as well as for 2.1% (n=7) of people had the minor significance.

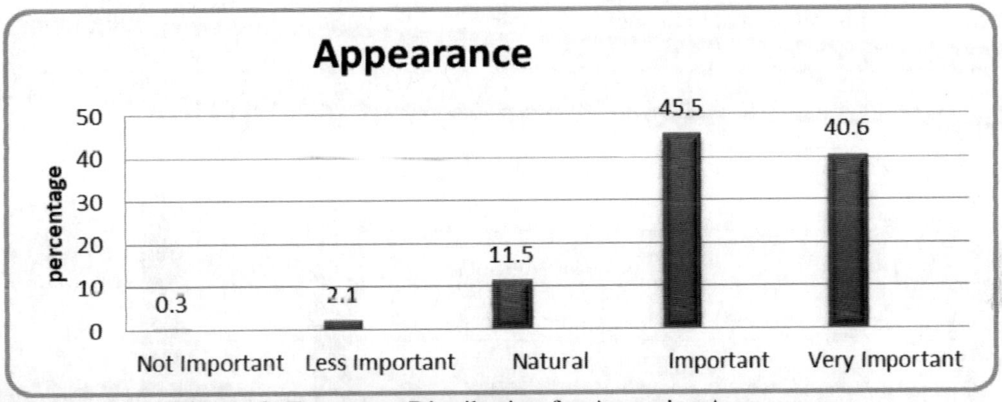

Figure 2. Frequency Distribution for Attractive Appearance
(Source: Field Survey, 2015)

The next question, which people have responded, people's expectation level toward the form of building in figure 3. For only one person(3%) was not important the form of building among participants, for 4.5%(n=150) of them had less significance as well. Moreover, the chart presents about one fifth of participants (n=68) did not give any idea about it. While for 39.4% (n=130) of them were very important the attractive form of building, for 35% (n=116) were less important.

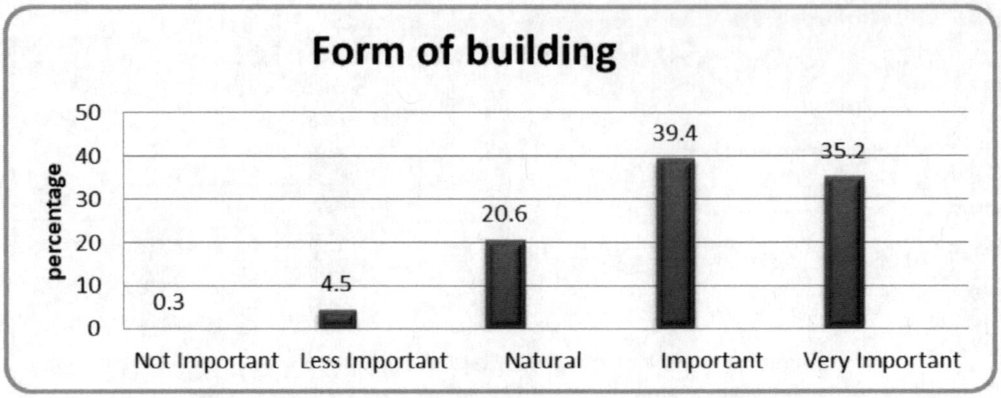

Figure 3. Frequency Distribution for Form of Building
(Source: Field Survey, 2015)

The figure 4 illustrates the expectancy of people toward good condition of ventilation in the community center. From the chart, it is clear that for more than half of people(58.8%,n=194) the suitable ventilation was very important matter. It is the single factor which most of the participants had claim very important among other factors in building condition of the community center. For 35.2% (n=116) of people the good ventilation in inner space was important. Whereas 5.25% (n=17) of participant didn't have any idea. Moreover, for 6% (n=2) of people the condition of ventilation and lighting had less importance on their expected level, for one person (0.3%) was not an important matter as well.

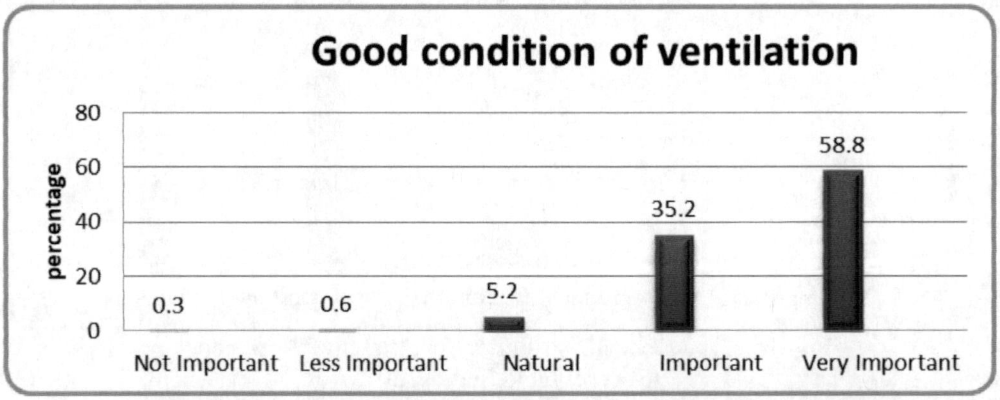

Figure 4. Frequency Distribution for Good Condition of Ventilation
(Source: Field Survey, 2015)

5. Conclusion

In terms of building design, for the majority of the participant's expected a good condition of the ventilation as the most importance factors compared to the others, as shown earlier in the analysis part. The quality of the building has straight relation to the achievement of the community center and a good social life is dependent upon a good building design. In community center; the floor space should have sufficient area to incorporate the vast of people, so the floor space and size of each space play significant role in community center. Another factor needs to consider is the influence of building microclimate. Air conditioning and its standards play a significant role in providing acceptable building in providing thermal comfort for the. Effects of lighting in a public building could, in principle, occur at multiple levels: the perceptual/cognitive level (i.e., the level of how different aspects of the space are perceived), the emotional level (i.e., how people affectively respond to the space) and the behavioural level (i.e., how participant behave in relation to the public building environment)(Quartier, Vanrie, & Van Cleempoel, 2014). Ventilation and lighting are two components which should be considered in the acceptable community center and they offer utilization of natural source can be a good strategy. The previous researchers presented arguments to emphasize the appearance of building especially in a public building. They mentioned that, physical attractiveness as an appearance consideration of a public building is one of the factors which influence the participant's welcome and affects its users. Each building plays significant role in making interesting and enjoyable spaces. For example, the fabric of the building, form of building should be attractive, flexible enough to encourage a variety of participation. Enhancing the building design can be one of important factors which should be considered in Malaysian community centers.

6. References

[1] Abidin, I. S. Z., Usman, I. M. S., Tahir, M. M., and Yap, Y. C. 2005. Characteristic of Attractive Square as Public Space : Putra Square , Putrajaya 4 Discussion : Case study Putra Square. *Journal of Energy, Environment, Sustainable Development and Landscaping*, 338–343.

[2] DBKL. 2004. *Draft of Kuala Lumpur Structural Plan 2020*. Kuala Lumpur City Hall.

[3] DBKL. 2014. *Draft of Community Center Spaces and Usage in Kuala Lumpur*. Kuala Lumpur City Hall.

[4] Gavin Tunstall.. 2006. *Managing the Building Design Process* (Second Edi., pp. 197–206). Elsevier Ltd.

[5] Glover, T. D. 2004. The "Community" Center and the Social Construction of Citizenship. Leisure Sciences, *26(1), 63–83. doi:10.1080/01490400490272486*

[6] Grodach, C. 2009. Art spaces, public space, and the link to community development. *Community Development Journal*, 45(4), 474–493. doi:10.1093/cdj/bsp018

[7] Jacobs, J.1961. The death and life of great American cities. *American Transactions on Engineering & Applied Sciences*, 10(3), 721–729

[8] Mess, H. A. and King, H. 1947. Community Centres and Community Associations, in H. A. Mess (ed.) Voluntary Social Services since 1918. *London: Kegan Paul, Trench, Trubner and Co. Community Development Journal*

[9] Nurul, W., Wan, M., and Rani, M. 2014. American Transactions on Engineering & Applied Sciences Understanding the Usage Pattern of Local Facilities in Urban Neighbourhood towards Creating a Livable City. *American Transactions on Engineering & Applied Sciences*, 3(2), 129 148.

[10] Quartier, K., Vanrie, J., & Van Cleempoel, K. 2014. As real as it gets: What role does lighting have on consumer's perception of atmosphere, emotions and behaviour? *Journal of Environmental Psychology*, 39, 32–39. doi:10.1016/j.jenvp.2014.04.005

[11] Smith, M. K. 2002. *Community centres (centers) and associations : their history , theory , development and practice*.

[12] Thomas, R. M. 2003. *Blending Qualitative and Quantitative Research Methods in Thesis and Dissertations*. United States: Crowin press.

[13] Utaberta, I. N., Nur, A., Goh, A., and Spalie, N. 2010. Evaluating Spatial Use And Design Development Of " Modern " Community Centre In Malaysia. *In "Empowering Modern Asian City Makers* (pp. 1–19).

[14] Whyte, H. W. 2000. How to turn a place around. *Projects for Public Space Inc*, 225-232.

[15] Worpole, K. 1992. *Towns for people: transforming urban life*. Open University Press.

Acknowledgments

This study was conducted with the aid of 'Geran Universiti Putra Malaysia 2017' (GP-IPS/2017/957730) for the research project "Critical analysis of Malay Traditional Architecture Values, Design and Thought in Malaysia Modern Built Environment" granted by Universiti Putra Malaysia. Authors also acknowledge the support of WARIS Research Group and Faculty of Design and Architecture (FRSB), Universiti Putra Malaysia.

IOP Conf. Series: Materials Science and Engineering **401** (2018) 012024 doi:10.1088/1757-899X/401/1/012024

Secondary school classrooms daylighting evaluation in Negeri Sembilan, Malaysia

R N Syaheeza[1], E M Husini[1], F Arabi[1], W N W Ismail[1,2] and M Z Kandar[3]

[1)]Architecture Department, Faculty of Engineering and Built Environment, Universiti Sains Islam Malaysia, Nilai, 71800, Negeri Sembilan
[2)]Center of Architecture and Built Environment Innovative (SeRAMBI), Programme Architecture, Secretariat of Identity of National Architecture, Faculty of Engineering and Built Environment, Universiti Kebangsaan Malaysia
[3)] Architecture Department, Faculty of Built and Environment, Universiti Teknologi Malaysia, Skudai, Johor

E-mail: rajanursyaheeza@gmail.com

Abstract. Natural daylight is a renewable energy source that is important in photobiology study which looks into the effect of light to living organism. Students' visual comfort satisfaction due to efficient daylight in a classroom are highlighted in the physical of the building and learning space such as windows and openings. One of the factor that influences the design of classrooms environmental condition. Environmental condition consideration includes the daylighting requirement for classrooms. Various standards and guidelines on daylighting consideration have been developed in Malaysia including the recommendation on Window-to-Floor Ratio (WFR) and illuminance level. However, the recommended WFR may not achieve the recommended illuminance level. This paper evaluates whether the recommended WFR design achieve the recommended acceptable illuminance level. Two classrooms in Kolej Permata Insan, Nilai, Negeri Sembilan with different orientation were selected for the study. Integrated Environmental Solution (IESVE) simulation software is used for daylight evaluation. The WFR of the two selected classrooms were set to the recommended 20%. Simulation results shows that the illuminance level for both classrooms are higher than recommended. The simulation also shows that the daylight distribution in the classrooms are not uniform, thus creating inefficient daylight condition. Therefore, the WFR recommends further studies with additional of daylighting design strategy considerations.

1. Introduction

Learning spaces design has evolved throughout the years due to various reasons such as construction technology, educational systems, government's policies, space requirement and environment condition. Throughout the classroom design development, a concept of 'open-air design' was popularized. This concept shifted the focus of classroom design towards daylighting considerations because the concept allows too much daylight amount in the classroom that causes visual discomfort to the students [1]. Various research proven that sufficient daylight in a classroom is required due to its positive effect towards the students including improvements on learning performance, alertness, cognitive skills, behavior, mood, circadian rhythm, dental strength and productivity [2, 3, 4]. Thus, various standards and guidelines were introduced all around the world.

Content from this work may be used under the terms of the Creative Commons Attribution 3.0 licence. Any further distribution of this work must maintain attribution to the author(s) and the title of the work, journal citation and DOI.
Published under licence by IOP Publishing Ltd

IOP Conf. Series: Materials Science and Engineering **401** (2018) 012024 doi:10.1088/1757-899X/401/1/012024

Malaysian Standard 1525 (MS1525) highlighted the recommended illuminance level for learning spaces, while Uniform Building by Law (UBBL) recommends the Window-to-Floor Ratio (WFR) for learning spaces. An interview with JKR personnel shows that the classroom design standards and guidelines still follows a standard introduced by British colonial government before Malaysia's independence [5] and may not suited to the recommended illuminance level that is currently being referred to by architects and designers. Therefore, this study will focus on the evaluation of daylighting performance in an existing classrooms of a public school in Malaysia. By identifying the daylighting performance of these classrooms via simulation software, it can show whether the recommended WFR can achieve the recommended illuminance level.

2. Literature Review

Efficient daylighting can improve students' optimal education performance such as visual comfort, subjective mood, cognitive performance, physical activities, sleep quality, attention and alertness in classroom [6, 7]. Furthermore, visual discomfort causes inefficiency in executing learning tasks, thus decreases the students' performance [8]. However, providing more daylight illumination level does not increase the visual acuity, but creates glare and visual discomfort as well [9]. Learning space that receive high daylight intensity with low glare, added with high quality artificial light source improves the visual comfort, which improves the learning space environment [10]. Proven that students learn to read 26% faster in a classroom with efficient daylight [11]..

The importance of daylight in classrooms are highlighted in various standards and guidelines. Illuminating Engineering Society of North America (IESNA) recommended 300 lux to 500 lux of illuminance level for learning tasks [12]. Others such as Malaysian Standard 1525 (MS1525), Public Works Department (JKR) and ZUMTOBEL Handbook recommends the same illuminance level of 300 lux to 500 lux for common reading task in general teaching spaces [13, 14]. Table I below shows the illuminance level recommended by few standards and guidelines. It Most of the recommendations are between 300 lux to 500 lux. Therefore, the recommended 300 lux to 500 lux will be the reference for the study.

Table 1. Illuminance level Recommendations.

Learning Space	Standards and Guidelines					
	IESNA	MS1525	JKR	ZUMTOBEL	EFM	OSHA
General Teaching Space	300-500	300-500	300-500	300	215-430	200
Science Laboratories	500	300	300			
Library	300	300-500	300			

Uniform Building by Law (UBBL) 2013, Section 39(3) stated that any room or space that serve the purpose of learning and education space should be designed with a minimum of 20% of Wall-to-Floor Ratio (WFR) for natural lighting and natural ventilation. Not less than 10% openings the WFR should be open-able for uninterrupted natural airflow [15]. A research shows that a public school in Ipoh, Perak that follows JKR design standards received recommended illuminance level [5]. Therefore recommendation of 20% WFR will be referred for this study.

3. Methodology

Measuring the illuminance level in an existing classroom requires the use of data loggers. Biesel (1953) introduced a method to measure illuminance level by considering the average, maximum and minimum illumination level of the space. These considerations are sufficient in measuring the illumination level [1], which is applicable for learning spaces such as classrooms. Measuring illuminance level at task level by luminance spot measurement on grid of 1m x 1m in the classroom [16, 17] was simulated to identify the uniformity of the daylight distribution during Malaysia intermediate sky condition [5].

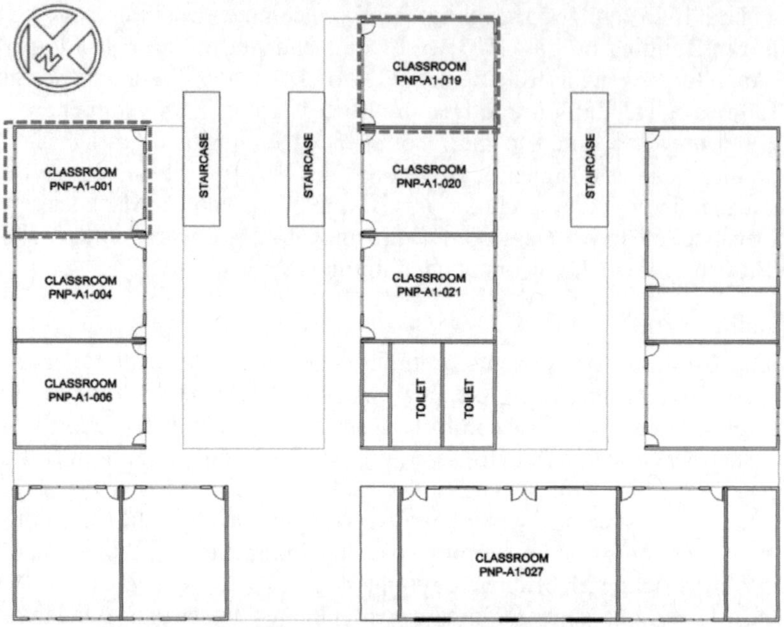

Figure 1. Kolej Permata Insan academic block ground floor plan.

Figure 1 above shows the two selected classrooms in Kolej Permata Insan (KPI), Universiti Sains Islam Malaysia (USIM) for the study. KPI is a secondary school operated since 2014. The school consists of academic block, administrative block, hostels and mosque. Classroom PNP-001 and PNP-019 are located on the ground floor of the academic block.

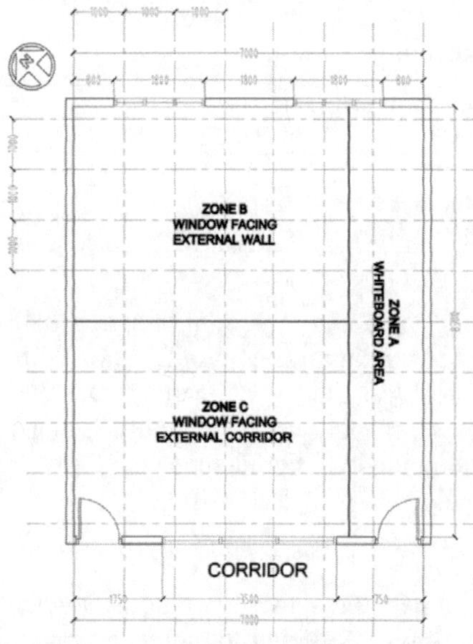

Figure 2. Classroom PNP-001 floor plan.

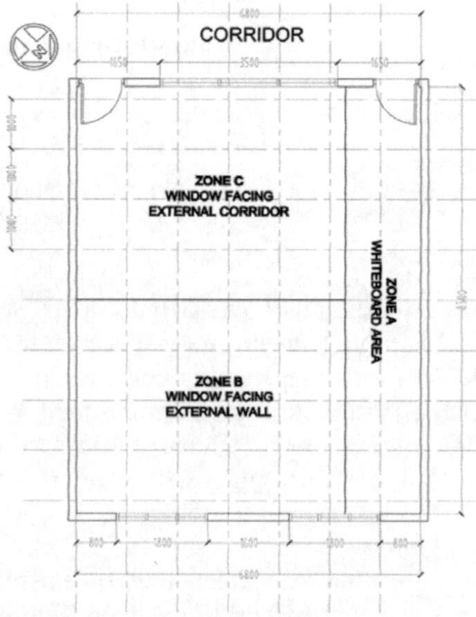

Figure 3. Classroom PNP-019 floor plan.

Figure 2 and 3 above shows the floor plan of the two selected classrooms. Both classrooms have similar layout design with windows on both side. Zone A is the whiteboard area, while Zone B are facing the outdoor and Zone C facing the corridor. The windows are located 900mm above the floor level. Classroom PNP-001 is facing southeast, meanwhile classroom PNP-019 is facing southwest.

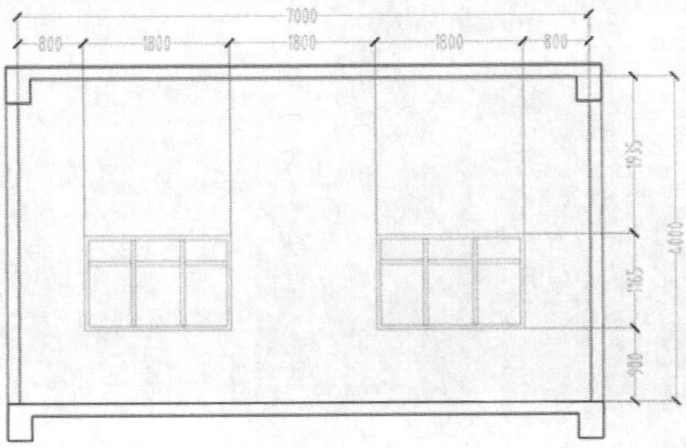

Figure 4. Zone B elevation.

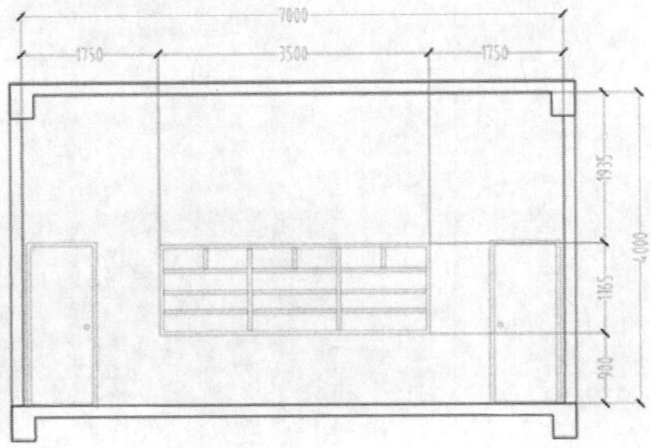

Figure 5. Zone C elevation.

Figure 3 shows the wall elevations for Zone B and Zone C respectively. Both side of the windows are facing the outdoor and the corridor respectively. No external shading device and internal blinds were installed. Glass windows on both side are positioned 900mm height from floor level and not glazed. The existing Wall-to-Floor Ratio (WFR) are between 14.5% to 15.5%, lower than recommended. However, the windows for simulation will be set at 20% of WFR recommended to test the illuminance level and daylight performance. Integration Environment Solution Virtual Environment (IESVE) is a software that is used in the study. The function Radiance in IESVe is a ray-tracing software to estimate daylight accurately.

4. Result and Discussion

Figure 4 below shows the result of the simulations based on recommended 20% WFR. The results show that both classrooms received average illuminance level higher than recommended, thus an inefficient daylighting. The highest illuminance level of 1413 lux and 1246 lux respectively occurs within 2m from both side of Zone B and Zone C windows. A few scattered point measurement shows recommended range of illuminance level, mainly 2m away from the windows. This shows that the daylight quality are not uniformly distributed.

WFR	PNP-001	PNP-019
20%	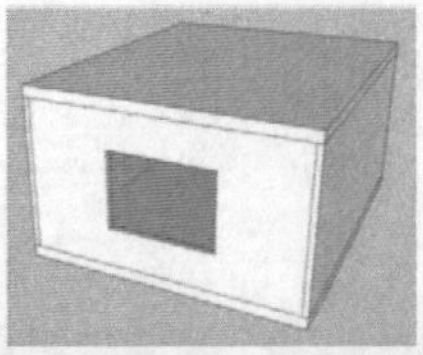 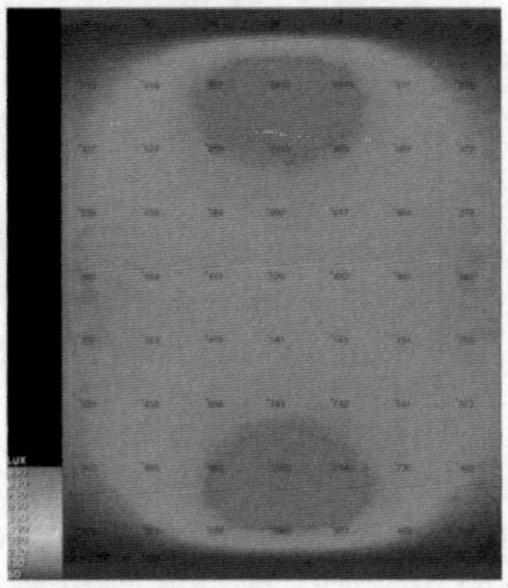 Ave. lux : 573.5 lux Min. lux : 129 lux Max. lux : 1413 lux	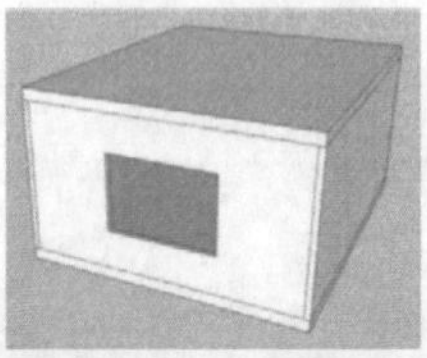 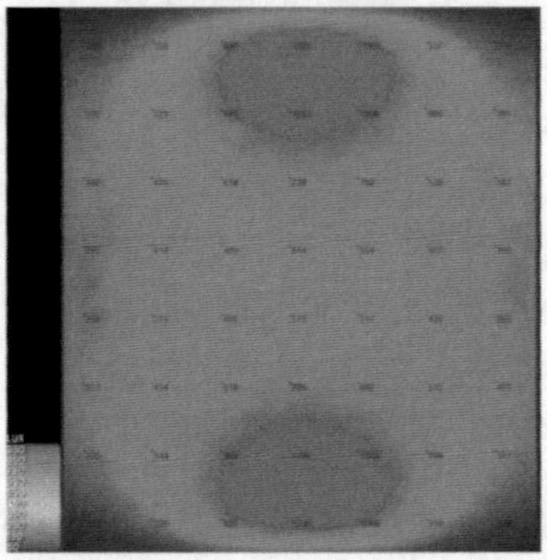 Ave. lux : 584 lux Min. lux : 132 lux Max. lux : 1246 lux

Figure 6. PNP-001 and PNP-019 simulation results.

5. Conclusion

The result of the simulations shows that independent from the orientation of the classroom, both received illuminance level higher than recommended by MS1525 of 300 lux to 500 lux, even though the WFR used were 20% as recommended by UBBL. Therefore, the recommended WFR should be furthered study to identify the suitable range so that the recommended illuminance level can be achieved. To achieve the recommended illuminance level in schools, architects and designers shall provide different designs due to the aspect of daylighting performance and the physical of the classroom such as the layout, orientation, and size of the opening.

Acknowledgement

This research is supported by the research Fundamental Research Grant Scheme (FRGS) 1/2015/TK07/USIM/02/1 and USIM Graduate Assistance Scheme.

References

[1] Wu W and Ng E 2003 A review of the development of daylighting in schools *Lighting Research Technology* **35** (2) 111-25

[2] Heschong L, Wright R L and Okura S 2002 Daylighting impacts on human performance in school *Journal Illumination Engineering Society* 101-14

[3] Samani S A and Samani S A 2012 The impact of indoor lighting on students' learning performance in learning environments: A knowledge internalization perspective *Int. Journal of Business and Social Science* **3** (24) 127-36

[4] Mirrahimi S, Ibrahim N L N and Surat M 2013 Estimation daylight to find simple formulate based on the ratio of window area to floor area rule of thumb for classroom in Malaysia *Research Journal of Applied Science Engineering and Technology* 931-5

[5] Mathalamuthu A D and Ibrahim N L N 2014 Evaluation of daylighting at public school classrooms in Ipoh, Perak *Alam Cipta* **7** (1) 27-34

[6] Shishegar N and Boubekri M 2016 Natural light and productivity: Analyzing the impacts of daylighting on students' and workers' health and alertness *Int. Journal of Advances in Chemical Engineering & Biological Sciences (IJACEBS)* **3** (1) 72-7

[7] Vi Le T H, Gillott M C and Rodrigues L T 2016 The Case for hybrid ventilated primary schools in Ho Chi Minh City in Vietnam *36th Int.Conf. on Passive and Low Energy Architecture* (Los Angeles: University of Nottingham)

[8] Gilavand A, Gilavand M and Gilavand S 2016 Investigating the impact of lighting educational spaces on learning and academic achievement of elementary students *Int. Journal of Pediatrics* **4** 360–9

[9] Lapus J A, *et al.* 2010 *Educational Facilities Manual* rev. ed Department of Education Philippines (Pasig City: Department of Education) p 224

[10] Barret P, Davies F, Zhang Y and Barret L 2015 The impact of classroom design on pupil's learning: Final results of a holistic, multi-level analysis *Building and Environment* **89** 118-33

[11] Heschong L, *et al.* 1999 *Daylighting in Schools: An Investigation into the Relationship Between Daylighting and Human Performance* Heschong Mahone Group (California: Pacific Gas and Electric Company)

[12] DiLaura D L 2011 *The Lighting Handbook: Reference and Application* Illuminating Engineering Society of North America (New York: IESNA)

[13] *Code of Practice on Energy Efficiency and the use of Renewable Energy for Non-residential Buildings* 2014 Malaysian Standard 1525:2014 2nd rev. (Malaysia: Department of Standards) p 24-31

[14] ZUMTOBEL 2018 *The Lighting Handbook* 6th ed. (Austria: Zumtobel Lighting GmbH) p 244

[15] *Uniform Building by-Laws* 2013 Malaysian Standard 16th ed. (Malaysia: Department of Standards) p 24-31

[16] Theodorson J 2009 Daylit classrooms at 47N, 117W insights from occupation *26th Conf. on Passive and Low Energy Architecture* PLEA2009

[17] Zomorodian Z S, Korsavi S S and Tahsildoost M 2016 The effect of window configuration on daylight performance in classrooms : A field and simulation study *Int. Journal Architecture, Engineering and Urban Planning* **26** (1) 15-24

IOP Conf. Series: Materials Science and Engineering **401** (2018) 012025 doi:10.1088/1757-899X/401/1/012025

Percentage Coverage of Tropical Climbing Plants of Green Facade

Mohd Khairul Azhar Mat Sulaiman[1*], Mohd Fairuz Shahidan[2], Maslina Jamil[1], Mohd Fauzi Mohd Zain[1]

[1] Centre for Innovative Architecture and Built Environment (SErAMBI), Faculty of Engineering and Built Environment, Universiti Kebangsaan Malaysia,Bangi, Selangor, Malaysia.
[2] Department of Landscape Architecture, Faculty of Design and Architecture, Universiti Putra Malaysia, Serdang, Selangor, Malaysia.
Email: m.khairulazhar@ukm.edu.my

Abstract. Green façade (GF) is one of vertical greenery systems used as passive solutions applied on building façade in improving indoor thermal environment and energy efficiency. However, optimum improvement of thermal environment from GF is dependent on the overall foliage coverage established on the system itself. Higher percentage coverage of the foliage is expected to regulate and lessen the heat transfer toward inside environment. Therefore, this study is to investigate the potential of selected tropical climbing plants in establishing foliage coverage on the GF. Three selected tropical species of climbers used on green façade; i.e. *Anemopaegma chamberlaynii* (AC), *Antigonon leptopus* (AL), and *Quisqualis indica* (QI). A small-scale experiment with small scale size of freestanding walls (1.5 m (height) × 1.0 m (width) × 0.15 m (thick)) attached with GFs (modular trellis) and planter boxes at ground were built for this study. Despite small size leaf, AC performed consistently and has the highest percentage coverage throughout the measurement period with average of 86.5 %. The dominant trait of AC which is vigorous and fast growth rate contributed to the dense and almost fully covered the trellis structure of green façade. Higher percentage coverage also estimated to generate an instantaneous effect on thermal environment modification.

1. Introduction

Wall and roof are an outer façade for the building to define the indoor spaces while providing security, privacy and protection. Also, the function included to offer optimum thermal environment for the end-users. However, this function sometime failed to be delivered by the building façade. Therefore, passive solution such as green façade (GF) is introduced and attached on building façade as shading device. GF is one of the vertical greenery systems (VGS) used widely to improve thermal conditions especially for indoor environment [1,2].

Performance of GF in improving indoor thermal environment has been widely studied by researchers. Coma et al. [2] studied an external surface temperature reduction of a one-room building or test cell wall with GF at the south orientation during summer in Lleida, Spain. Maximum reduction obtained was 14 K. The study also investigated on the differences of internal surface and indoor air temperatures of the test cell. The maximum results were 2 K and 1 K, respectively. Using the same research tool, Price [3] conducted an experiment using one-room building to explore the potentials of IGF on cooling effect performance in Clarksville, USA. He attached IGF at the south and west facing aspects of the test cell.

Content from this work may be used under the terms of the Creative Commons Attribution 3.0 licence. Any further distribution of this work must maintain attribution to the author(s) and the title of the work, journal citation and DOI.
Published under licence by IOP Publishing Ltd

One test cell was treated as control room. The temperatures of western and southern aspect walls with GF were decreased at a maximum of 11.3 K and 6.4 K, respectively. The indoor air temperature of the test cell with GF near to the west wall was 1.8 K, meanwhile the south wall recorded 1.0 K. At the west orientation of the indoor environment, the difference of heat transferred was 10.7 W/m^2 and at the south, the difference was 3.6 W/m^2.

Another full-scale setup experiment by Koyama et al. [1] was conducted to investigate shading and transpiration cooling effect on three identical test cells, with one was left as a control unit. Two of the test cells produced lower indoor air temperature, with maximum peaks at 3.3 and 4.4 K compared to the control unit. The maximum difference for external surface temperature was 9.5 K. The experiment was then continued with climbing plants on GF at one of two test cells were severed, and another one was left intact to calculate the transpiration cooling effect. The cooling effect performance of severed plants was decreased. The indoor air and internal surface temperatures of the wall with an intact plant of GF were 0.4 K and 1.2 K lower than a severed plant test cell.

Sunakorn and Yimprayoon [4] piloted a full-scale building on empirical investigation of GF in Thailand. The experiment was run on the existing identical rooms. The study was designed to explore the use of GF with the influences of natural ventilation. The results revealed that the differences in room air temperature of IGF attached in front of openable windows were lower than control room. Without natural ventilation, the IGF room temperature was 0.5 K lower than the non-covered room. With natural ventilation allowed to go through the room, the indoor temperature was 0.9 K lower.

From the all above mentioned studies, the foliage coverage established on the GF is an important factor in determining the successfulness of this system. Koyama et al. [5] found strong positive relationship between cooling effect performance of GF and coverage percentage (r: 0.815). Higher percentage of foliage coverage contributed to higher wall surface temperature reduction. Therefore, this study is to investigate the potential of selected tropical climbing plants in establishing foliage coverage on the GF. This is an initiative to establish fundamental information of tropical climbing plants coverage performance used on GF.

2. Material and Methodology

2.1. Plant Selection

Plant selection for this experiment was based on several criteria suited for this experiment. Several criteria have been considered for this experiment in order to expect the best performance of coverage and density of foliage on the GF. The criteria include the growth rate, coverage, weather tolerance, maintenance, climbing pattern, leaf size, and availability in the nursery. Most importantly, the plant must be a native climbing plant of tropical biomes. An indigenous plant owns by surrounding climatic condition promotes better growth development and coverage performance. Native plants of tropical regions are mostly evergreen all through the year. Located near to the equator, vegetation receives abundant solar radiation throughout the year. Thus, the results of measurement were assumed can be predicted for an annual performance.

Taking into deliberation the harsh conditions of the experimental site and the duration of the study, the preferred climbing plant should have the characteristic of vigorous and fast of growth rate of plant, spreading coverage pattern that can cover GF within the time frame given, full sun plant and tolerate to the extreme weather conditions, low maintenance, twining and clinging climbing pattern, large leaf size, and high supply at the nursery. The summary of the selection criteria is given in Table 1. In addition, popular climbing plants used on GF at Malaysia's building were also considered to help the architect or landscape architect to have essential information on the plants that are applied for their building and for the future.

Reviewed through literature, observed through current situation of GF in Malaysia's construction industry, and contacted with local landscape architects and gardeners, this study had identified three species potentially to be investigated. The potential species are *Anemopaegmea chamberlaynii* (AC),

IOP Conf. Series: Materials Science and Engineering **401** (2018) 012025 doi:10.1088/1757-899X/401/1/012025

Antigonon leptopus (AL), and *Quisqualis indica* (QI) (Figure 1). The criteria of selected plants are tabulated in Table 2.

Table 1. Summary of selection criteria for climbing plants to be used for this experiment

Criteria	Selection Criteria	Remarks
Growth rate	Vigorous and fast	Limitation of research time for plant to grow
Coverage	Spreading	Able to maximise coverage on supporting structure within time frame given
Weather tolerance	Full sun	Full sun to act as shading device and able to withstand extreme weather conditions
Maintenance	Low	Low on fertilizing, pesticides monitoring and watering
Climbing pattern	Twinning and clinging	System selected with trellis support structure, not suitable for adhesive type climbers
Leaf size	Large	Larger leaf provides larger shading coverage
Availability in nursery	High supply of preferable climber	Limited supply in the market
Popularity	Most used in the construction industry	For better understanding on plant performance that been used in the industry

(a) (b) (c)

Figure 1. Specimen of selected tropical climbers (a) *Anemopaegma chamberlaynii*, (b) *Antigonon leptopus*, and (c) *Quisqualis indica*.

Table 2. Summary of potential criteria of selected climbing plants

Criteria	*Anemopaegma chamberlaynii*	*Antigonon leptopus*	*Quisqualis indica*
Growth rate	Vigorous	Vigorous	Fast
Coverage	Spreading	Spreading	Cascading down
Weather tolerance	Full sun and hardy	Full sun and semi hardy	Full sun and hardy
Leaf size	Small – 5cm	Medium -13cm	Medium-15cm
Climbing pattern	Twinning with hooked tendrils	Tendrils	Stem twinning
Maintenance	Low	Low	Medium

IOP Conf. Series: Materials Science and Engineering **401** (2018) 012025 doi:10.1088/1757-899X/401/1/012025

2.2. Small-scale Experiment

The small-scale experiment also used by Jim and He [6], Koyama et al. [5], Laopanitchakul et al. [7], and Wong et al. [8]. Three mock-ups of GF with free standing walls were built for this experiment. The size of freestanding walls were 1.5 m (height) × 1.0 m (width) × 0.15 m (thick) and trellises were 1.5 m (height) × 1.0 m (width) with white planter boxes at the bottom. Figure 2 shows detailed measurement of GF mock-up. Each planter box was planted with three numbers of climbers from each species. The heights of each of the climbers were almost similar before they were transferred into respective planter boxes. Climbers were left for a month for the establishment of foliage before any measurements were taken. Modules were arranged at west facing orientation. The experiment was conducted for duration of 30 days period on 10 September - 9 October 2012 at Faculty of Design and Architecture, Universiti Putra Malaysia (2.99°N, 101.71°E).

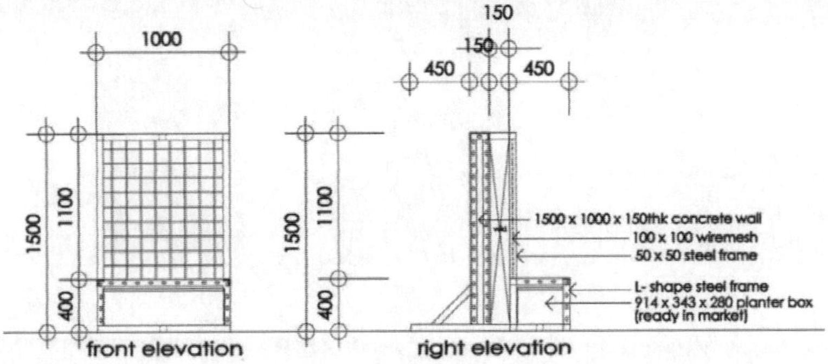

Figure 2. Details measurement of small-scale of IGF

2.3. Percentage Coverage

Average of coverage performance of the climbing plants was determined by the percentage coverage on selected three days (first, middle and final days of the experiment. The percentage coverage was the percentage area of the behind wall covered by the plants to the area of the wall. The percentage of the coverage was calculated through binary images. The binary images were obtained from converted digital images taken by digital camera (Nikon D3100). The images were converted to binary image and the calculation of shaded area was adopted from pixel segmentation feature in Adobe Photoshop CS4. An example of the image conversion is shown in Figure 3.

(a) (b)

Figure 3. Image conversion of *Quisqualis indica* from (a) digital image to (b) binary image

3. Results and Discussion

Percentage coverage for three species is demonstrated in Figure 4 and tabulated in Table 3. The average performance of AC, AL and QI is 86.5 %, 79.6 %, and 69.4 %, respectively. In the overall performance, despite of small leaf size, AC performed significantly remarkable. Throughout the experiment, the development in the percentages coverage of AC are almost stagnant, however, the percentage coverage is significant, with percentage ranges of 85.9 to 86.9 %. Vigorous growth rate of AC established dense foliage coverage is expected to provide instantaneous thermal environment improvement. The performances of AL and QI declined towards the end of the experiment. Both of the species performed exceptionally at the beginning of the experiment, however, the decrement of percentage is suspected to relate with the leaf lifespan and the pattern of foliage coverage.

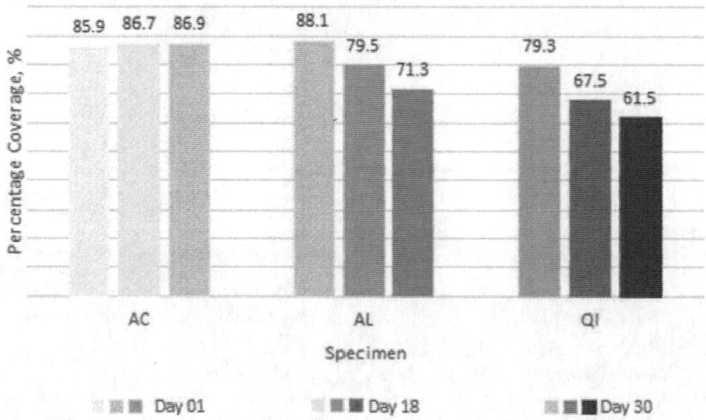

Figure 4. Percentage coverage of *Anemopaegma chamberlaynii* (AC), *Antigonon leptopus* (AL) and *Quisqualis indica* (QI) for day 1, 18 and 30 of the duration of conducted experiment (10 September – 09 October 2012).

Table 3. Percentage coverage of *Anemopaegma chamberlaynii* (AC), *Antigonon leptopus* (AL) and *Quisqualis indica* (QI) for day 1, 18 and 30 and the average of the duration of small-scale experiment (30 days; 10 September – 09 October 2012). Area of coverage is calculated in m^2 over area of trellis of IGF (1.5 m^2).

Species	Day 01		Day 18		Day 30		Average	
	%	m^2	%	m^2	%	m^2	%	m^2
Anemopaegma chamberlaynii (AC)	85.9	1.29	86.7	1.30	86.9	1.30	86.5	1.30
Antigonon leptopus (AL)	88.1	1.32	79.5	1.19	71.3	1.07	79.6	1.19
Quisqualis indica (QI)	79.3	1.19	67.5	1.01	61.5	0.92	69.4	1.04

The performance of QI is considered as the least percentage performed from all the specimens. The lessened performance towards the last day of the experiment is because of the pattern of foliage coverage. The foliage pattern of QI is cascading down. The stem climbs twinning vertically over the supporting structure with less number of leaf coverage before reaches the highest point of the structure. Only at the top position of the trellis, foliage starts to form the cascading down pattern. This pattern only establishes maximum coverage when reaching the top point of supporting structure such as trellis and take longer times to establish the optimum coverage. Thus, it is explained the reason for the weak performance of QI even though on the small coverage area. Nevertheless, the percentage coverage of QI is still acceptable for the given timeframe of the experiment.

Meanwhile, the diminution performance of AL is caused of short leaf lifespan. From the observation, the leaf tended to grow vigorously, then the leaf started to experience fast rate by wilting, drying and dying. The abundance of solar radiation in the tropics increases the rate of photosynthesis assimilation of the leaf. Therefore, the longevity of the leaf is shorter which suspected due to the high rate of photosynthesis assimilation especially for the tropical plants [9].

The selection of the plant to be applied on GF is a very important decision to be taken. Larger size of leaf can contribute to higher percentage coverage. Nevertheless, the performance of AC is remarkable with an average of 86.5 % of 30 days experiment, despite small leaf size (~5 cm). The high growth rate of AC increases vigorously the leaf layer to form the dense foliage by overlapping each other in 3- and 4-leaf layers. Therefore, the coverage pattern and growth rate of the climbing plant trait are criteria to be prioritised. However, further study on more species and on relationship between plant traits and cooling effect performance of tropical climbers need to be undertaken.

4. Conclusion

Performance of selected tropical climbing plants in providing higher percentage of foliage coverage on GF is significantly remarkable. Percentage coverages for all species are above 50 % throughout the experiment timeframe. AC is performed significantly compared to other climbers. Coverage pattern and growth rate are two important traits to be prioritised in the plant selection. More species of tropical climbers should be conducted in further study, including relationship between plant morphology and cooling effect performance.

Acknowledgement

The authors would like to acknowledge the Universiti Kebangsaan Malaysia (UKM) and Ministry of Higher Education for supporting this research through Fundamental Research Grant Scheme (FRGS/1/2015/SS111/UKM/02/2).

References

[1] Koyama, T., Yoshinaga, M., Maeda, K. and Yamauchi, A. 2015. *Ecological Engineering* **85** 343-353.

[2] Coma, J., Pérez, G., Solé, C., Castell, A. and Cabeza, L.F. 2014. *Energy Procedia* **57** 1851-1859.

[3] Price, W.J. 2010. *Green façade energetics*. (Unpublished Master Thesis. University of Maryland, USA.)

[4] Sunakorn, P. and Yimprayoon, C. 2011. *Procedia Engineering* **21** 34-41.

[5] Koyama, T., Yoshinaga, M., Hayashi, H. Maeda, K. and Yamauchi, A. 2013. *Building and Environment* **66** 96-103.

[6] Jim, C.Y. and He, H. 2011. *Ecological Engineering* **37** 1112-1122.

[7] Laopanitchakul, V., Sunakorn, P. and Srisutapan, A. 2008. *International Conference Sustainable Building 2007, Seoul, Korea*. 26-27 June 2007.

[8] Wong, N.Y., Tan, A.Y.K., Chen, Y., Sekar, K., Tan, P.Y., Chan, D., Chiang, K. and Wong, N.C. 2010. *Building and Environment* **45** 663-672.

[9] Kikuzawa, K. and Ackerly, D. 1999. *Plant Species Biology* **14** 39-46

IOP Conf. Series: Materials Science and Engineering **401** (2018) 012026 doi:10.1088/1757-899X/401/1/012026

A study of the classification of crowd types to build crowd-sourcing-based GIS

Jung, Seo-young[1], Yu, Jung-ho[2]*

[1] PhD. Student, Department of Architecture Engineering, Kwangwoon University, South Korea

[2] Professor, Department of Architecture Engineering, Kwangwoon University, South Korea

*Corresponding author E-mail: myazure@kw.ac.kr

Abstract. The Smart City concept is a city model that aims to solve city's problems and improve the life qualities by combining new advanced technologies such as ICT and big data. The Smart City concept concentrates on humanity, where complexed technologies are applied in a city, and makes a community of collaborating human life and technologies. In this concept, there are continuous efforts to solve many social problems such as safety and welfare for people. Therefore, it is imperative to improve the method for people to participate in solving social problems; the crowd sourcing technology, which uses crowd mapping and community mapping, is a noticeably improving method. In this study, we review the theoretical background of crowd mapping through the foundation study to make a crowd mapping model and classify several crowd types in the case of disabled welfare systems.

1. Introduction

The Smart City concept is a city model that aims to solve city problems and improve the life qualities by combining new advanced technologies such as ICT and big data. This concept implements 4th modern technologies such as Artificial Intelligence (AI), Internet of Things (IoT) and Cloud Computing. This concept also gains attention as the core power of innovated developments and a space where various technologies are combined.

South Korea has been proceeding the U-City industrial project from 2005. The concept of the U-City focused on technology in South Korea. However, a Smart City concentrates on humanity, where complexed technologies are applied in a city, and makes a community of collaborating human life and technologies. In this concept, there are continuous efforts to solve many social problems such as safety and welfare for people. Therefore, it is imperative to improve the method for people to participate in solving city problems; the crowd sourcing technology, which uses crowd mapping and community mapping, is a noticeably improving method.

Crowd mapping is the combination of PPGIS (Public Participation GIS) and crowd sourcing. The PPGIS defines the people who use the internet, and the GIS defines the process of people's participation. People can attend the spatial decision-making process by making a map, which is analysed like normal maps and GIS. Crowd sourcing is a compounded term of "crowd" and "out-sourcing". In short, people can easily access to the entire process to make a map and participate in the spatial decision-making process; hence, it is "community (crowd) participatory mapping".

To advance information and communication technologies, the implementation of the E-government is the main objective, and the opinions to let people have a chance to participate in the social and spatial

Content from this work may be used under the terms of the Creative Commons Attribution 3.0 licence. Any further distribution of this work must maintain attribution to the author(s) and the title of the work, journal citation and DOI.
Published under licence by IOP Publishing Ltd

IOP Conf. Series: Materials Science and Engineering **401** (2018) 012026 doi:10.1088/1757-899X/401/1/012026

decision-making process are strengthening. In this situation, crowd mapping, which has been guiding people's participation in, can be applied as a solution to several social problems about the decision making of spatial policies and maximize the effectiveness of the GIS.

Thus, in this dissertation, we review the theoretical background of crowd mapping through a foundation study to make a crowd mapping model and classify several crowd types in the case of disabled welfare systems.

2. Participatory GIS and the Concept of Crowd Mapping

2.1. Concept of crowd mapping

Crowd mapping originates from the concept of crowd sourcing, which is a foundation of the internet cloud circumstance. Crowd sourcing is a compounded term of "crowd" and "out-sourcing", which means the solution is obtained from people's participation to solve their social problems. The crowd sourcing concept is based on the analysis and sharing of information, which is naturally formed from big data.

Through crowd mapping, people can learn about interesting local places near them and participate in various processes and communications from the planning stage to the decision-making stage.

Crowd mapping is related to big data and cloud system from the ICT viewpoint. Big data, formal or informal, are obtained from methodologies, tools, saved files, search results, analyses, and visualizations. With increasing usage of smartphones and navigation systems, the Location-Based Service (LBS) is remarkable nowadays, and geographical information is separated from the existing big-data concept and classified as Big Geo Data. Big Geo Data have a limitation because they are provided by the existing big-data technology. To efficiently handle big geo data, cloud technology has emerged as an alternative. In cloud computing technology, the combined internet service can access to portable devices with no telecommunication barriers.

2.2. Classification of Crowd Types in Crowd Mapping

Heipke (2010) divided the people's level of the proportion of geographic information by crowd sourcing (table 1).

Table 1. Crowd types (Source. Heipke, 2009 / Anne Marthe Ruitton-Allinieu, 2011)

Crowd type	**Example**	**Motivation**	**Contribution**	**Data**	**How to get them involved**
Groupies	Retired GIS experts	Map lovers Public/Community acknowledgement	Continuous	Trustable & very valuable Little effort, large value	Community forums
Casual users	Hikers, bikers, mountaineers, etc.	For personal use	Occasional	Little effort, little value	Moderated community platforms

Experts	Active and leading map experts in organizations (mountain rescue, fire brigades, civil protection, etc.)	For personal use To simplify their work	Continuous	Very valuable and trustable	Financial agreements
Media mappers	Once-off mappers activated by media campaigns	Through fun activities (competitions, mapping parties, etc.)	Limited in time and extent. Depends on the frequency of the campaigns	Large effort	Try to turn them into casual users
Open mappers	Users of OpenStreetMap, Google MapMaker, etc.	Public recognition Using good public data		Very valuable and large	Develop open-source mapping tool
Passive mappers	GPS-enabled mobile phones' owners	Money		Unsure about how to use it	Still restricted by law
Mechanical Turks		Money		Must be carefully checked	Publishing HITs

3. Classification of Crowd Types in Korea

It is imperative that the users can conveniently access the information. The users can be sorted into several levels based on the geographical information. It is essential to crowd map according to the user's level.

Crowd mapping can be classified as follows:

- Administrator of public geographical information: The Administrators, who are experts of geographic information and producers of the original information source who perform maintenance and management tasks. Because South Korea has begun to construct GIS following domestic policies, GIS has a public feature. Thus, the administrator's role occupies large proportions of GIS management. According to the area of disabled welfare, social workers and governors of the social department are considered administrators.
- Supplier of commercial geographic information: The suppliers who work with commercial geographic information related to the market of crowd mapping.
- User of geographical information: The user uses this information for smartphone maps, map service on internet and navigation information. In terms of disabled welfare, the disabled who need help and volunteers are considered the user group.

- Positive participants of geographical information: These people are active users, provide current information and revise the errors. Presently, the number of crowd mapping participants has been increasing with the use of smartphones. For disabled welfare, the coordinators who organize volunteering programs are considered positive participants.
- Participants of potential usage: These people unintentionally use and produce geographical information. The advert's information is notably important in space information big data. The normal users unintentionally use social media such as Twitter and search keywords; this information can be meaningful if the geographical features are linked with geographical information. These people appear to be the basic source of geographical information.

4. Discussion

In this study, we reviewed the theoretical background of crowd mapping through the foundation study to make a crowd mapping model and classified several crowd types in the case of disabled welfare systems. The results of this study will be used as a basis to create a crowd-sourcing-based smart map for the disabled to advance smart city services in Korea.

References

[1] Anne-Marthe Ruitton-Allinieu, 2011, Crowdsourcing of Geoinformation: data quality and possible applications, Master's Thesis, Aalto University School of Engineering Department of Surveying.

[2] ChungWeon, Oh., 2013, A Study on Volunteered Geographic Information System in Internet Environment, Journal of the Korean Cartographic Association 13(1), p.127-141

[3] Goodchild, M.F., 2012, The Quality of Big (geo)Data, Big Data in Geographic Information Science Panel 2012, http://stko.geog.ucsb.edu/bigdatagiscience2012.

[4] Heipke, C., 2010, Crowdsourcing geospatial data,ISPRS Journal of Photogrammetry and Remote Sensing, 65(2010),p.550-557

[5] Hristova, D., Mashhadi, A., Quattrone, G., and Capra, L., 2012, Mapping Community Engagement with Urban Crowd-Sourcing, Sixth International AAAI Conference on Weblogs and Social Media.

[6] Parker, B., 2006, Constructing community through maps? Power and praxis in community mapping, The Professional Geographer, 58(4), 470-484.

[7] Rowe, G. and Frewer, L.J., 2000, Public participation methods: A framework for evaluation, Science, Technology, & Human Values, 25(1), 3-29.

[8] Sieber, R., 2006, Public participation and geographic information systems: A literature review and framework, Annals of the American Association of Geographers, 96(3), 491-507.

Acknowledgments

This work was supported by the National Research Foundation of Korea (NRF) grant funded by the Korean government (MSIT) (NRF-2017R1A2B4012228)

IOP Conf. Series: Materials Science and Engineering **401** (2018) 012027 doi:10.1088/1757-899X/401/1/012027

Identification of external variables for the Technology Acceptance Model(TAM) in the assessment of BIM application for mobile devices

S H Hong[1] and J H Yu[2]

[1] PhD. Student, Department of Architecture Engineering, Kwangwoon University, South Korea
[2] Professor, Department of Architecture Engineering, Kwangwoon University, South Korea
E-mail : myazure@kw.ac.kr

Abstract. With the development of information and communication technology (ICT), various tools designed to utilize the ubiquity of portable devices have been launched. This is also true within the construction industry, where many such tools seek to enhance work efficiency by utilizing ICT and to transform PC-based environments into mobile-based environments to enhance the convenience of use. In accordance with this trend, building information modeling (BIM) has now been extended to portable devices. A number of BIM tools are currently available, such as Formit (provided by Autodesk), BIMx (Graphisoft) and BIMsight (Tekla); however, they are limited to model-related functions such as viewer and model modification. Recently, additional functions such as communication and cloud-based information sharing have been developed, but user-oriented demand research using tools are not yet available. Therefore, based on a literature review, this study summarizes the external variables which affect the use of BIM application tools and describes the expected effects when the relationships between these variables are hypothesized and verified using the technology acceptance model. This study contributes to the development of user-oriented BIM utilization tools for portable devices.

1. Introduction

The use of portable devices in the construction industry has become increasingly common due to continuous advances in information communication technology (ICT; Lee et al., 2015). In particular, building information modeling (BIM) has been the focus of various functional tools based on portable devices. Originally, BIM software such as Revit (provided by Autodesk) and ArchiCAD (provided by Graphisoft) was designed to be used on personal computers. However, to extend usage to mobile devices, a diverse range of applications has appeared (e.g., Formit by Autodesk, BIMx by Graphisoft, and BIMsight by Tekla).

These new programs have typically focused on revising architectural models and displaying the results for viewers. Information sharing based on cloud computing and support services for communication have been added to these applications, but an in-depth analysis of user needs for each of these functions has not yet occurred.

A common tool employed in user need surveys is the technology acceptance model (TAM). The main variables of concern when developing a system is time to market, flexibility, integration, and user satisfaction (Kim, 2015). The TAM defines external variables that can affect user behavior and it can

Content from this work may be used under the terms of the Creative Commons Attribution 3.0 licence. Any further distribution of this work must maintain attribution to the author(s) and the title of the work, journal citation and DOI.
Published under licence by IOP Publishing Ltd

be have a definition of the requirements of user's view for making system. The TAM has been employed in a variety of areas, including shopping, travel, sports, and the analysis of the organizational acceptance of BIM (Lee et al., 2013). Based on this, it would be efficient demand survey even if it is applicated to making BIM tool based on portable devices.

Therefore, the present study conducts a literature review of research related to the development of portable devices in order to analyze the requirements for BIM mobile tools based on the TAM. It also provides the process in which about to external variables for making BIM mobile tools based on time to market, flexibility, and integration. This study thus provides suggestions regarding the requirements for the development of BIM tools and can act as a foundation for further research on this topic.

2. The Identification of External Variables for the TAM

The TAM stems from the theory of reasoned action (TRA). It utilizes the constructs "perceived usefulness" and "perceived ease of use", which are individual beliefs that are affected by external variables. Both of these constructs, in turn, affect "attitude toward using" and "behavioral intentional use" (David, 1989). The TAM is used to represent the mutual relationship between external variables which affect the acceptance of technology by a user and factors which affect actual behavior. This model hypothesizes a relationship between external variables and both perceived usefulness and perceived ease of use. However, because the TAM is founded on individual beliefs, it has a limitation in that social influence is ignored.

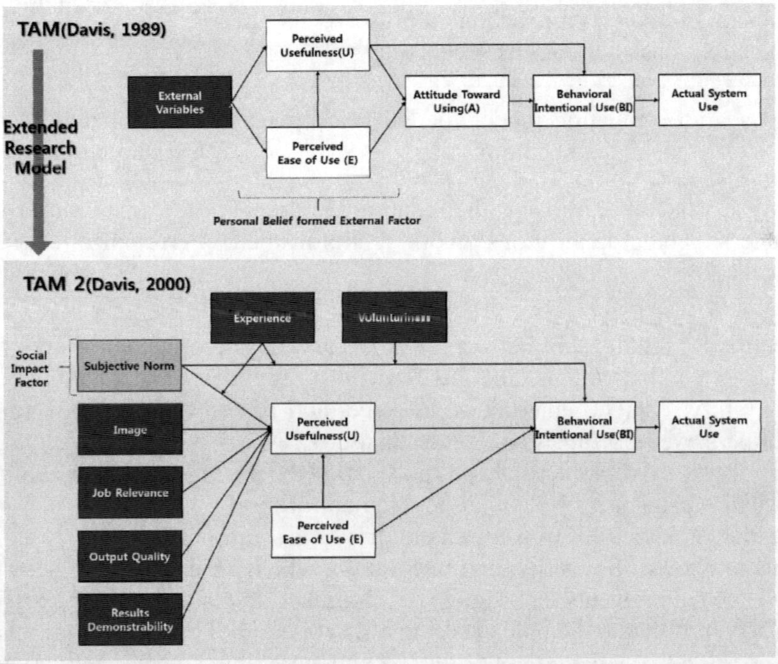

Figure 1. Davis's Technology Acceptance Model(TAM)s (1998,2000)

Due to this limitation, Venkatesh and Davis (2000) introduced the TAM 2 (or ETAM) as an extension of the original TAM. In the TAM 2, attitude is excluded as a variable and subjective norms are added, which is related to social influence, it had ever been deleted (Park, 2012). The TAM 2's external variables are distinguished to both the social influence process and identical tool process. The social process includes subjective norms, voluntariness, and image, while the identical tool process includes job relevance, output quality, and result demonstrability (Song et al., 2011). More recently, the TAM 3 has been proposed, suggesting additional external variables which may affect perceived usefulness

(Venkatesh & Davis, 2008). TAM has been referred to over 4,000 times by various researchers (Lee, 2014).

The development of tools for portable devices is predicated on the following advantages: 1) the possibility of an internet connection anytime and anywhere, 2) the ease of communication, and 3) the convenience of portability. Based on this, a number of studies have sought to improve user acceptance and ultimately increase the usage of applications by applying the TAM. In a preliminary study, perceived usefulness and perceived ease of use are separated and set as external variables which affect the content of the application. These studies thus define the factors that are likely to influence the content of the application. If the relationship between the elements is positive, the external factor influences the content of the application, but if the relationship between the elements is negative, the external factor does not influence the content. In addition, they also provided a structural diagram of the TAM.

Table 1. External Variables Used in the TAM

Author (year)	Content	External Variables
Park (2012)	Examines the relationships between concepts related to a travel agency application and describes implications for the use of the application	Quality of Content, Faith of Application, Travel Company
Kim (2012)	Analyzes the motivation behind the use of mobile applications based on four external factors	Utility, Quality of Content, Entertainment, Price, Design
Lee (2014)	Proposes an evaluation model for BIM acceptance based on organizational culture	Personalization, Organization, Collaboration Environment
Shin (2016)	Analyzes factors affecting the continued use of mobile shopping applications among adults in their 20s and 30s.	Interaction, Information Offer, Personalization, Playfulness, Instant Connectivity

Park (2012) identified a relationship between positive and negative factors in applying the extended TAM to a travel agency's application. Kim (2012) reviewed commercial applications and employed five external factors (utility, content, entertainment, price, and design) in the TAM to investigate their relationship with perceived usefulness and ease of use. Lee (2014) defined external variables which affect organization culture or BIM acceptance and verified the hypothesis. Lastly, Shin (2016) analyzed the level of acceptance using the TAM for a shopping application.

The external variables identified from the literature review are presented in Table 1 and classified according to time to market, flexibility, and integration, which are factors that need to be considered when establishing the requirements for program development. According to Kim (2015), time to market enables users to accomplish very complex tasks in a short period of time. Flexibility is the ability of an application to work effectively regardless of the working environment, and integration means it can be used in conjunction with existing systems.

The external variables considered in the literature review can be divided into six types. Time to market is associated with external variables related to the convenience of providing information, such as utility, quality of content, and information offer. Flexibility is associated with external variables related to the connection with the environment, such as personalization and instant connectivity. Integration relates to variables related to the convenience of communication and information sharing, such as a collaborative environment and interaction. In addition, entertainment and playfulness are related to playability, design is related to convenience of use, and price is related to economic efficiency.

Therefore, the external variables that should be considered in the TAM model for the development of BIM application tools for portable devices are the convenience of providing information (time to market), the connection environment (flexibility), communication and sharing of information (integration), playability, convenience of application use, and economic efficiency. If the TAM can be verified by setting hypotheses related to perceived usefulness, perceived ease of use, attitude toward using, and behavioral intentional use, it is expected that BIM application tools for portable devices can be developed to reflect user acceptance.

3. Conclusions

With the development of ICT, the development of portable device-based technology has taken on increased importance, with the use of ICT also increasing in the construction industry. If BIM application tools for portable devices that manage building information from the PC environment are developed, the convenience and universality of technology use are expected to improve. However, although various portable device-based technologies have been developed, it is seldom the case that their functions are implemented based on a needs analysis of potential users.

Therefore, in this study, external variables for the TAM are identified based on a literature review in order to develop BIM applications for portable devices. This study analyzes past research that has used the TAM to develop portable device-based applications in other industries and examines the external variables used in those studies. Based on this, external variables that can be used to develop BIM application tools for portable devices are categorized into six types.

This study does not attempt to verify the hypothesized relationships using the TAM, but rather identifies external variables that can be used in the TAM. Therefore, in future research, external variables for the six categories should be specified based on the factors suggested in this study, and the hypotheses verified by establishing the relationships between the external factors. This study contributes to fundamental research regarding the development of application functions that reflect user requirements.

Acknowledge

This research was supported by a grant(17CTAP-C114926-02) from Infrastructure and transportation technology promotion Program funded by Ministry of Land, Infrastructure and Transport of Korean government.

This research was supported by a grant (18AUDP-B127891-02) from the Architecture & Urban Development Research Program funded by the Ministry of Land, Infrastructure and Transport of the Korean government.

Reference

[1] Park L W 2012 Utilization and Circulation of Smart Phone Mobile Application of Travel Agencies through Expansion of the Technology Acceptance Model(TAM), Kyung-Hee Univ., Philosophy of Doctor

[2] Cho J H 2015 Extended technology acceptance model(ETAM) through acceptance of sports applications, Chung-Ang Univ., Master course

[3] Shin M A 2016 The Effect of Mobile Shopping Application Characteristics on Continuous Use Intention : Focused on Technology Acceptance Model, Inha Univ., Master Course

[4] Kim S S 2012 An Empirical Study on Users' Intention to Use Mobile Applications, Soongsil Univ., Philosophy of Doctor

[5] Song Y H and Hur W M 2011 Exploratory Study of Adoption and Diffusion of Premium Digital Convergence Product: Moderating Effecting of Social Value, Korea Knowledge Management Society 12(1), pp53-76

IOP Conf. Series: Materials Science and Engineering **401** (2018) 012027 doi:10.1088/1757-899X/401/1/012027

[6] Lee S K Yu J H and Jung D 2015 BIM Acceptance Model in Construction Organizations, Journal of Management in Engineering, 31(3)
[7] Lee S K 2014 BIM Acceptance Readiness Evaluation Model considering Organization Culture, Kwangwoon Univ., Philosophy of Doctor
[8] Lee H S, Lee J K, 2015 A Study on the BIM (Building Information Modeling) – compatible Applications based on the Analysis of Web and App as an Architectural Design Tool in Mobile Environment, Journal of Korea Design Knowledge 33, pp143-152
[9] Kim C S 2015 Easy Way to Learn the Software Engineering, Han-bit Academy

THERMAL COMFORT STUDIES ON HOUSES IN HOT ARID CLIMATES

Noor Aziah Mohd Ariffin, Amira Behaz, Zuraini Denan

Dept. of Architecture, International Islamic University Malaysia

Abstract. The traditional houses in Ghardaia City in Algeria had been built using traditional construction materials and passive solar design (PSD) strategies for the hot desert climate. These thousand-year-old houses are still standing and conditioning the harsh desert climate while maintaining indoor thermal comfort for its occupants. Of late, two trends of modern housing and construction were observed to have evolved in Ghardaia. Firstly, were the modern houses (Mod.H) following the international style and secondly, the eco-friendly houses (Eco.H), modern houses but utilising passive solar design. The Mod.H was observed to implement some of the traditional features but using the modern construction material. Meanwhile, the Eco.H modified the characteristics of the traditional construction and materials carefully in a modern context. Analyses were conducted to investigate the thermal performances of the three house designs. The intention was to ascertain the best construction that can achieve and maintain thermal comfort according to ANSI-ASHRAE 2014, in this harsh desert climate. Based on the Trad.H model, comfort data the two more modern houses were also monitored. Data collection was carried out in July, which was one of the hottest months. The study indicated that passive solar techniques employed in the Trad.H were able to maintain the comfortable indoor environment for the occupants without any mechanical aid. The indoor temperature of the Eco.H was 3 to 9 °C higher, and the Mod.H difference was 5 to 9°C above the thermal comfort zone. The study ascertained that designs of buildings with traditional construction and material played a significant role in moderating the indoor environment when the outdoor climate of the Algerian desert was unbearable. Inadvertently, energy for cooling was saved creating a significant impact not only on the economy both at the national and global level, but on the environmental scale as well.

Keywords: Thermal comfort, Passive Design Strategies, Ghardaia, Traditional houses, Eco-friendly houses, Modern houses, hot desert climate.

1. Introduction

The issues of thermal comfort have become a global point of interest to many researchers around the world. This fact is attributed to the realisation that people are spending most of their time (more than 90%) in an indoor environment (Lee & Chang, 2000). Passive methods of achieving thermal comfort in buildings in several countries have been commonly investigated (Sharples & Malama, 1997). These passive methods are the solution to provide the comfort of the indoor environment, healthy lifestyle and energy efficiency (Hegde & Ananthakrishna, 2008). Thermal comfort is considered an essential and necessary aspect in the urban sector especially in the desert climate where the outdoor temperature reaches 50oC in the summer season and freezing point during winter. Diurnal swings of 15-20oC or more are common with minima often reaching below 15oC. Under these climatic conditions, the ability to store

Content from this work may be used under the terms of the Creative Commons Attribution 3.0 licence. Any further distribution of this work must maintain attribution to the author(s) and the title of the work, journal citation and DOI.
Published under licence by IOP Publishing Ltd

energy is a vital strategy for controlling and improving the building microclimate and lowering energy demand weighing on the infrastructure. The thermal mass strategy is highly recommended to maintain thermal comfort in this regions, and this is the characteristic of the unique fortified human settlement of the Algerian desert (Chabi & Dahli, 2014).

This paper attempted to look for an explanation of the little known "Ksar" (plural is Ksour), a fortified human settlement in the desert - on its traditional design and passive solar design in Ghardaia city in the middle of Algeria. Houses are packed tightly together to protect against the extreme climate, and these designs have stood the test of time for over a thousand years (O.P.V.M Ghardaia, 2014). However, in recent years modern house designs and construction were on the increase compromising the traditional designs. In the interest of thermal comfort, the paper compares the thermal performances of these unique houses. The first was a traditional house (Trad.H) build around the 11th century. The other two houses: the modern house (Mod.H) and the ecological house (Eco.H) were more recently developed, after 2000.

2. Aim and Objectives

This research aimed to investigate the thermal performances of the two contemporary houses about the Trad.H's. The objectives were to determine whether the Eco.H or the Mod.H with contemporary designs provided better indoor thermal comfort in the context of the summer season. Another objective was whether these houses required any improvements to be taken as a good model for the future construction in Algerian desert cities.

3. Ghardaia at a Glance

Ghardaia is located in the centre of the northern part of the Algerian desert. It is situated 600 km south of the capital Algiers between 32° Northern latitude and 2°30' Eastern longitude. The climate of the Ghardaia city is a hot desert climate characterised by summers with torrid heats reaching 50°C and soft winters with the average minimum just above freezing point. Relative humidity is very low except for the winter months where 60% is common. Precipitation is low and less than 25 mm throughout the year. The thermal comfort zone (TCZ) for the hot and arid climate of Algeria as according to ASHRAE 55-2014 standards is 18°C < T < 30°C.

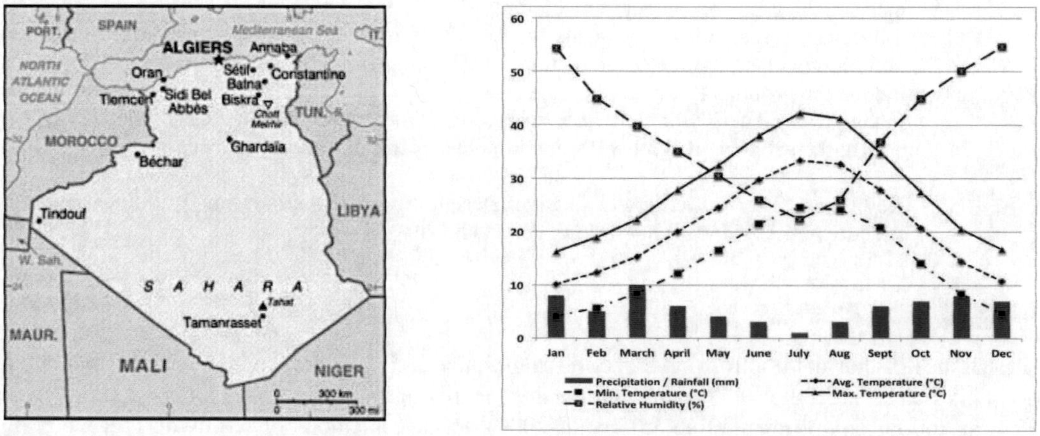

Figure 1: The location and climate of Ghardaia, Algeria
Source: Britannica.com

Due to the intense heat in the dry desert the desert cities in Algeria and also North Africa are defined by a grouped habitat called *"Ksar"*. The fortified morphology of the *Ksar* reiterated the historical need to protect itself from external threats and to cope with climatic conditions. The *Ksar* is an agglomeration of

IOP Conf. Series: Materials Science and Engineering **401** (2018) 012028 doi:10.1088/1757-899X/401/1/012028

very tight houses usually surrounding a mosque in the center and fortified. The number of houses in one *Ksar* may exceed 1000 houses (Bisson, 1957). Some literature relates that once the population growth exceeds the capacities of the mosque, it is necessary to build another *Ksar* and a new city around it, as shown in Figure 2.

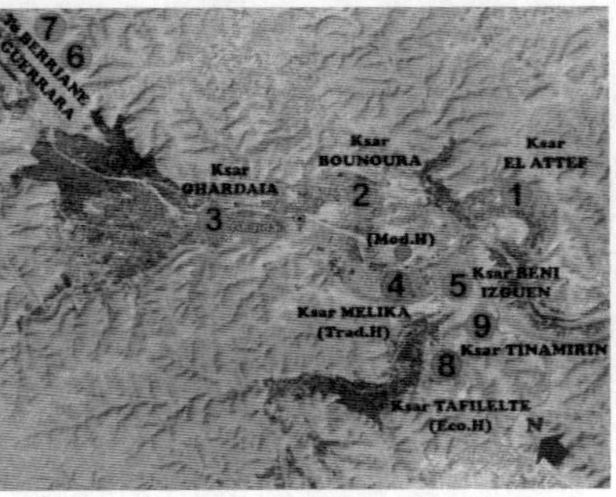

a) Locations of *Ksour* in Ghardaia

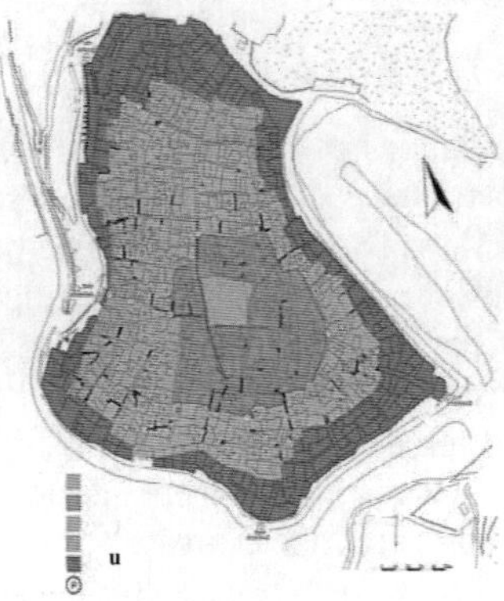

b) *Ksar* Melika

c) Packed arrangement of houses in a *Ksar*

Figure 2: *Ksar/Ksour* in Ghardaia city, Algeria

4. Passive Design Strategies (Pds) and Thermal Comfort

In achieving thermal comfort in the hot desert climate, Szokolay (2014) recommended several passive design strategies. Amongst them are the use of thermal mass to condition the internal environments. Thermal mass is equivalent to thermal capacitance or heat capacity, i.e., the ability of a body to store thermal energy. When outside temperatures are fluctuating throughout the day, a large thermal mass within the insulated portion of a house can serve to moderate the daily temperature fluctuations. The traditional design of the houses in a *Ksar* uses this PDS to its fullest. Not only were the houses built with stones and clay at least 500mm thick for the envelope, as far as possible only the entrance façade is exposed, if not shaded by neighbouring buildings across narrow alleys.

5. Traditional House Construction

The traditional case-study house (Trad.H) in *Ksar* Melika was designed based on this PDS principle. The house consisted of 3 to 4 rooms grouped around an open court called Wast-eddar (centre of the house). It is the central space that organises the dwelling and the primary means of daylight into the house within the densely packed residential of a *Ksar* (Chabi & Dahli, 2014). By its form and position, it regulates the organisation of the whole dwelling, as well as its functioning. *Wast-eddar* is the most active space in the house and in addition to its roles, it ensures aeration and lighting throughout the home by the skylight grill called the *'Chebek'*. All other main rooms (entrance, living, dining and kitchen are in direct relation with *Wast-eddar*. A superposition of the *Wast-eddar* has the effect of reducing the radiant heat inside the house in the summer, by covering it with cloth or canvas. As for the functional top terrace on the roof, it was reserved for women and used for sleeping at night. The terrace was a flat and heavy slab, allowing the reduction of heat transfer into the house (Ali-Toudert, et al., 2005). Other climatic integration strategies characterised by the Ghardaian houses started with a south orientation to benefit the oblique sun rays in winter and the vertical rays in summer (Ali-Toudert, et al., 2005). Stones were the most used building material and characterised by a high thermal capacity. The stones have the advantage of capturing solar energy and accumulating it, to restore it later and easily evacuated at night by natural ventilation effect.

Of late modern houses built in Ghardaia do not follow the traditional guidelines. As a result, these houses became energy intensive to achieve and maintain thermal comfort in the harsh climate in the desert of Algeria. It had become urgent to intervene in the design and urban planning of Ghardaia because of the increasing level of demographic growth and energy usage. This paper investigated the thermal performances of the three different types of houses to provide scientific data of the bioclimatic requirements essential for the construction of these houses.

6. Methods

To understand the thermal performances of the various houses of the *Ksour* the methods undertaken in this study were the buildings measurement survey and climatic data monitoring. The building survey was pertinent for this study to get similar physical and climatic parameters for consistency of collected data and results. Three houses were selected due to their differing house types as required for this study and from different *Ksour*. The first is the Trad.H from *Ksar Melika*, the second a typical Mod.H *(not in any Ksar*) and the third a modern house but built with ecological principles (Eco.H) from *Ksar Taffilelt*. The three houses were selected, and physical measurements were taken and drawn on Archicad. The environmental parameters used for the analysis were indoor air temperature, relative humidity and the outdoor air temperature. The readings are taken during summer on various days in July 2017. Measuring instruments and devices include Onset data logger to measure relative humidity and air temperature continuously and the infra-red thermometer to measure the surface temperature.

Building descriptions

Modern house (Mod.H)

The Mod.H was constructed in 2000 and located approximately 1.5 km from the Trad.H. The layout of the buildings is different than the closely-packed Trad.H. This house is 73 sq.m. area and was constructed using 400 mm thick stone external walls in the basement level, 300 mm concrete external walls in upper floors and 100 mm thick internal walls of hollow concrete blocks. The materials used for the ceiling were the concrete compression slab of reinforced concrete beams and 200 mm thick reinforced concrete roof. The internal height of the rooms is 2.6 m. It has the most common layout in present day modem houses.

Eco-friendly house (Eco.H)

The Eco.H, also constructed in 2000, was located in *Ksar Tafilelt* with a built-up area is 96 sq. m. The construction materials used were stone, concrete and plaster. Stone exterior walls of 400 mm thick, constituted the building-supporting structure of the house as well as all the walls on the façade. The internal walls were made of hollow concrete blocks 150 mm thick. The materials used for the ceiling were the concrete compression slab of reinforced concrete beams, spaced 650 mm apart. The plaster vaults ensured the thermal and sound insulation. The void between the compression slab and the vaults was filled with a mixture of lime and sand which were also used in the exterior cladding to allow better malleability of the mortar.

7. Results and Discussion

Data loggers were located at five different locations in the houses. But this paper discussed those situated in the living, kitchen and Wast-eddar areas only. In all three houses, only the kitchen walls were proximal to the outdoors. The other walls of the houses were attached to neighbouring buildings, if at all.

Table 1 summarises the results of indoor temperature monitoring and thermal comfort analyses in the three spaces of the case study houses.

Table 1: Results of data monitoring

Living	Min °C	Max °C	Diurnal °C	Diff. with TCZ (K)	Diff. outdoor/indoor (K)
Trad.H	30.4	30.6	0.2	0.4 - 0.6	22
Mod.H	31	32.1	1.1	1 - 2	10.8
Eco.H	30.1	32.3	2.2	0.1 - 2	11.2
Kitchen					
Trad.H	30.4	30.6	0.2	0.4 - 0.6	22
Mod.H	34.9	37.2	2.3	5 - 7	16
Eco.H	31.9	35	3.1	2 - 5	18
Wast-eddar					
Trad.H	33.7	34.4	0.7	4 - 5	22
Mod.H	37.2	38.9	1.1	7 - 9	16
Eco.H	36.8	38.7	1.9	7 - 9	18
Outdoor	35	54.6	20	5-25	

Data monitoring was conducted on a sweltering summer day on 15/7/2017 where the maximum temperature was almost 55°C, and the minimum was 35°C with a diurnal range of nearly 20 K. Although all indoor temperatures monitored were above the upper limit of the thermal comfort zone (TCZ) of 30°C, all houses were able to moderate the indoor climate below the maximum outdoor temperatures.

In all three spaces monitored in the houses, the Trad.H performed the best where the difference of the level of the thermal comfort zone of only 0.4 to 0.6 K. Only in the wast-eddar the thermal condition there was 4 – 5 K above the comfort level of 30°C because the wast-eddar is the most exposed space in the house. During the period the survey it was noticed, however, that only a fan was used, ensuring energy efficiency. The fan was required to alleviate the stagnant hot air. The Mod.H fared the worst of all three spaces monitored. It was gathered that the thickness and material of the walls had the least thermal capacity between the three houses resulted in the higher temperature variations. The kitchen and the wast-eddar spaces were between 5 and 9 K above the TCZ. Five air-conditioners and eight fans were frequently used in this house. The wast-eddar in the Eco.H was the hottest space in the house with a difference of 7 to 9 K above the comfort zone. The roof construction is similar to Mod.H. Air-conditioning was required to be switched on to maintain comfort. In all three spaces monitored, the living rooms of the three houses were able to moderate the indoor temperatures closest to the upper limit of the comfort zone. This was because the living rooms do not have openings to the outside.

Relative humidity (RH) data was also monitored, and results showed that when outdoor conditions were so dry (ave. 10%) the indoors of the Trad.H, Mod.H and Eco.H were 25%, 24% and 20% respectively. Although still on the low side, the RH indoors were more comfortable. Energy consumption was also recorded, and the Eco.H used twice and the Mod.H thrice more than the Trad.H. From the results, it is apparent that the traditional house retained the temperature nearest to the comfort zone as indicated in the ASHRAE 55-2014 standards. The Eco.H can provide a better thermal comfort than the Mod.H, within the dense surrounding environment at Ghardaia city.

8. Conclusion

Results from data monitoring taken during the hot summer month of July showed that the indoor temperatures indoors of all three houses were much lower than the outdoor temperature. The Trad.H performed the best to retain the temperature nearest to the comfort range as indicated in the ASHRAE 55-2014 standards. The Trad.H still has better ability in providing the indoor thermal comfort for the more extended periods than the two modern houses in the same condition. Low thermal conductivity due to less thickness of walls and using non-local building materials like concrete in the Mod.H showed the higher temperature variations compared to Trad.H and Eco.H. The ultimate aim of providing thermally comfortable and energy efficient building is thus achieved by incorporating passive design strategies with traditional construction materials. This plays a significant role to save energy creating a significant impact on the economy. The study ascertained that designs of buildings with traditional construction and materials played a significant role in moderating the indoor environment when the outside climate of the Algerian desert was unbearable. For a more holistic understanding of the thermal performance of Ghardaian houses, a study during the cold winter months is pertinent.

Acknowledgement

This research was part of the FRGS Grant from the Ministry of Higher Education, Malaysia (MOHE) under the purview of the International Islamic University Malaysia (IIUM).

References

[1] Ali-Toudert, F., Djenane, M., Bensalem, R. and Mayer, H. (2005). Outdoor thermal comfort in the old desert city of Beni-Isguen, Algeria. *Climate Research*, Vol. 28.
[2] ASHRAE Standard 55-2014, (2014). Thermal Environmental Conditions for Human Occupancy. American Society of Heating, Refrigerating and Air- Conditioning Engineers, Atlanta, GA.
[3] Bisson, J. (1957), The Gourara Study of Human Geography. *IRS Brief No.3*. University of Algiers.
[4] Chabi. M., and Dahli, M. (2014). Heritage: A referent for urban renewal? Case of ksour of M'Zab. (1347). 1-13.
[5] Hegde, L.S., and Ananthakrishna, K.S. (2008). Indoor temperature in vernacular, conventional and alternative technology construction - a comparative investigation. *J. Indian Instit. Archit.*, 16–18.
[6] Lee, S., and Chang, M. (2000). Indoor and outdoor air quality investigation at schools in Hong Kong. Chemosphere, 41(1), 109-113.
[7] O.P.V.M Ghardaia. (2014). Office of Protection and Promotion of the M'zab Valley. www.opvm.dz.
[8] Sharples, S. & Malama, A. (1997). Thermal performance of traditional housing in the cool season in Zambia. *Renew. Energ*. 8, 190–193.
[9] Szokolay, S. (2014). Introduction to Architectural Science. Architectural Press Elsevier, UK.

IOP Conf. Series: Materials Science and Engineering **401** (2018) 012029 doi:10.1088/1757-899X/401/1/012029

Green Open Spaces Criteria to Achieve Social Interaction of Karkh Community in the City Baghdad, Iraq

S A Salih[1] and S Ismail[2]

[1]Doctor of Philosophy Student at Department of Architecture, Faculty of Design and Architecture, Universiti Putra Malaysia, Jalan UPM, 43400 Serdang, Selangor, Malaysia.
[2]Lecturer at Department of Architecture, Faculty of Design and Architecture, Universiti Putra Malaysia, Jalan UPM, 43400 Serdang, Selangor, Malaysia.

E-mail Address: sarah_alsaadi1990@yaho.com, sumarni@upm.edu.my

Abstract. This paper outlines the issue on open spaces, which led to decrease social interaction among residents in Baghdad city nowadays. The main objective of the paper is to identify the criteria of green open spaces to achieve sound social interaction in Baghdad city, Iraq. This paper employed quantitative method, in the form of survey, for data collection. Data were obtained from questionnaires, through the selection of 270 respondents in a single-stage random procedure from ten specific neighbourhoods in *Karkh* district. The study findings confirm that the appropriate criteria of open spaces and parks is essential to enhance social interaction. The results of this study are useful reference for urban and landscape planners, architects, social psychologists, the Municipality of Baghdad, and researchers in this field.

1. Introduction

Social interaction can be described as an interchangeable sequence of dynamic exchanges through which individuals can attach meaning, interpret, and respond [17]. While public space allow people to meet on ostensibly neutral grounds in planned and unplanned ways, to interact with others within the context of the whole community. Social relationships vary according to the age, gender and place, where place element is a key factor to determine the nature of social activity for all ages [12]. Factors affecting social interaction in public spaces can include; activities, spaces design, maintenance, management, safety and security [12]. Therefore, physical space is an essential to achieve social interaction, so place and its elements could be a key factor to solve social interaction issues [12].

Baghdad is the capital of Iraq and largest city in Iraq. Located along Tigris River which runs through Baghdad center and divided it to two parts *Karkh* the western shore of the Tigris and Rusafa the eastern shore of the Tigris [2]. Since the establishment of Baghdad by the Abbasid caliph *Al-Mansur* (766 CE) it had become a cultural, historical and social center for Arab and Islamic civilization. Baghdad was characterized by gardens, orchards and cultural and social diversity [1-2]. Presently, the City of Baghdad has been exposed to multiple changes as result of political conflicts, religious diversity, ethnic diversity, administrative changes and technological developments. All of these factors lead to disintegration and change in social interaction, neglecting many recreational facilities and open spaces in Baghdad [11].

Content from this work may be used under the terms of the Creative Commons Attribution 3.0 licence. Any further distribution of this work must maintain attribution to the author(s) and the title of the work, journal citation and DOI.
Published under licence by IOP Publishing Ltd

IOP Conf. Series: Materials Science and Engineering **401** (2018) 012029 doi:10.1088/1757-899X/401/1/012029

This paper elaborates the issues of lacking of social interaction as a result of lacking of an appropriate criteria of the green open spaces in the city of Baghdad. Therefore this paper employed quantitative method by survey for data collection from 270 respondents in ten specific neighbourhoods in *Karkh* district, Baghdad. The objective of this paper based on paper issue is to determine the criteria of a recreational green open space, to promote social interaction in *Karkh*, Baghdad.

2. Social Interaction in Green Open Spaces

The design of outdoor spaces has a great role to lead individuals to outside spaces, socializing and being together, also space design tools can be used to provide social interaction and the feeling of being a community, especially in transformation places [22]. Complex social diseases can be cured through different house surroundings, parks, and new spatial arrangements [22]. To achieve social interaction physical space must be provided, thus parks have crucial roles to develop and enhance cities as well as social relations, making them not only places where people join the nature but also communicate with each other socially and culturally [15]. While, to achieve a successful public park, good of access and linkage (GAL) should be the main factor in designing a park, followed by sociability (SOC), user and activities (UAC) and degree of comfort and image factors (DCI) [21]. Park was an important space in the relationship of man and nature to promote and provide space for physical activity, health behaviour and can reduce some diseases such as diabetes and certain cancer [21]. Parks also provide the benefits of community interrelationship, as well as increase the value of the property [21].

The development of social relationships in an urban scale should begin with careful attention to specific spaces by implementing the use of interaction design approaches [24]. Specific spaces should be used to effectively address the sociocultural relationships of the people [24]. With the shift to a service-based society, providing opportunities for outdoor recreation that enable mental and physiological self-regulation has become an increasingly important landscape function [25]. The quality of the nearby recreation area is at least as important condition for these benefits as the easy access to these areas. Inhabitants' satisfaction with the recreation areas, their activity and interaction level within the recreation areas, as well as the time spent in the recreation areas appeared to be more relevant predictors for these benefits than the frequency of visits in these areas [25].

3. Social Life, Recreation and Landscape in Baghdad, Iraq

Since the Baghdad foundation, it was given a special interest for recreational facilities, open spaces, green areas, and parks. While, Baghdad's name has been accompanied with the names of paradise, garden, and orchard. The tributary of Tigris flowed in all districts in Baghdad, to promote its beauty and splendour, and the vast orchards extended along the banks of this tributary [3]. While, the first zoo was established in 797, during the *Harun Al-Rashid* rule, which contained various types of birds, rabbits, fishes, monkeys, lions, etc. [3]. Some open spaces were also used to play golf or horsemanship [3]. While social life of Baghdad in Abbasid period (762-1258) based on two different social layers; 1) caliphs and ministers life lived a comfort and luxury life, and 2) local people who lived a simple life with high social interaction, where some poor people shared the same house [2]. Community of Baghdad was diverse in ethnic and religious, but it was open and amicable therefore, e.g. many of Muslim men married a Christian women [2]. People in Baghdad shared the important social events e.g. Friday prayer, Eid, Ramadan, promenade after Eid, weddings and festivals and etc. [2]. The period of founding the first Iraqi state in 1920 had a great influence on Iraqi society, where political awareness appeared, attention to follow the news and newspapers, emergence the movement of constitutional claim, referendum appeared for the first time in Iraq [7-8]. The revolution has also influenced a lot in the culture of Iraqi society, it became more coexistence community after the revolution [5-6-7-8].

IOP Conf. Series: Materials Science and Engineering **401** (2018) 012029 doi:10.1088/1757-899X/401/1/012029

War of 2003 in Iraq has led to loss of lives, social capital and destroyed the Iraqi infrastructure. This substantially lowered the quality of life, led to the inability to provide essential services and rendered state-building activities even more difficult [13]. Ethnic relations in Iraq have been strained since the last war 2003, ethnic relations have been particularly uncooperative and violent. The crisis aggravation between different ethnic in Iraq led to the decay the interaction between them [20]. Since 2003, the establishment of military-controlled zones, lengthy strips of T-walls, road-blocks and checkpoints interrupted movement along the arterial roads linking the different areas of Baghdad [14]. These barriers have created a widespread congestion and made routine journeys longer, meaning that local residents prefer to shop, work and socialize within their neighbourhood without interacting with other neighbouring people [14]. As well as many local open spaces where families would commonly walked or children played, have become dumping sites for garbage or collection areas for sewerage and stagnant water [11]. Recreational places, such as city parks and stadium and activities such as walking along or boating on the Tigris were either closed or off limits since 2003 [11]. The lack of safe public spaces means that families have little respite from the pressures of daily life [11-20].

In 1960 In 2007

Figure 1. Umma Park after and before the War of 2003
Source: algardenia.com

Existing parks in Baghdad do not perform their role effectively, this in turn reflected on residents' interaction which limited to indoor spaces only [19]. Responsible authorities should make more efforts for open spaces and parks maintenance, construction and administration, as well as provide a variety of facilities especially children's facilities could improve the function of this spaces as well as provide more parks in Baghdad could improve the interaction of community [19]. Governmental efforts toward this issue are also slight as a result of political and economic issue experienced by the country since the war of 2003, also charities and the general financial allocations only belong to humanitarian crisis and living conditions [19].

4. Paper Methodology

This study was implemented to identify the basic criteria of green open spaces to promote social interaction of *Karkh* communities, Baghdad, Iraq. *Karkh* district in Baghdad was chosen as a research area. Where *Karkh* is the western half of Baghdad, its area is around 2,650 km2, it consist of about ten basic sections (neighbourhoods) according to Ministry of Planning, Baghdad. The ten neighbourhoods are characterized by physical, social, economic, educational and cultural variances. The ten neighbourhoods are; *Amiriya*, *Mansour*, *Jihad*, *Ghazaliya*, *Mamoun*, *Saidiyah*, *Salehia*, *Doura*, *Baiyaa*, and *Utaifiyya*. The population of these ten neighbourhoods was about one million person, according to the Ministry of Planning, Baghdad in 2009. While, the respondents selected by using a simplified formula [23] as shown below. Where n is the sample size, N is the population size (1,000,000), and e is the level of precision (±6%). So the respondents were 270 people, taken in a

IOP Conf. Series: Materials Science and Engineering **401** (2018) 012029 doi:10.1088/1757-899X/401/1/012029

single-stage, random procedure from the specific neighbourhoods, and these 270 respondents is representing the whole district (*Karkh* district). The individual characteristics of respondents varied in terms of age groups (18-55), two genders, various income levels, education, occupation and accommodation.

$$n = \frac{N}{1 + N(e)^2}$$

Equation 1. A Simplified Formula for Proportions
Source: Yamane, 1967

This paper adapted Creswell recommendations to design the methodology of this study. A quantitative method was adapted in this study in terms of a survey by using questionnaires to assess the population opinion about the appropriate criteria of open spaces and parks to enhance social interaction. Closed-ended questions used as a basic tool to collect data, where (Edith, Joop and Don) international handbook of survey researches used to design the questionnaires of the study. Then information evaluated using statistical analysis by Statistical Package for the Social Sciences (SPSS) version 23.

5. Analysis and Discussion

5.1. Sample Characteristics

The 270 respondents were taken from different locations within the neighbourhoods like universities, work sites, streets, and shops. Regarding the sample gender, female respondents made up (51.9%) of the total sample which is (4.5%) more than male respondents. Male respondents made up (47.4%) of total sample, and there are two missing answers for gender survey. All respondents were found to fall within the age range from 18 to 55 years, where most of respondents were between the ages of 18 to 24 years (41.9%). While only a few respondents within the ages between 47 to 55 years (6.3%), as it was very difficult to get respondents within this age group from the selected places for the survey (Table 1).

The number of respondents from each neighbourhood was almost close, where the largest respondent number (12.2%) was taken from *Amiriyah*, while the lowest respondents' number (7.4%) was taken from *Utaifiyya*. As for the income of respondents, the majority (81.1%) represented middle income level, and (15.2%) of the respondents were high income level and only (3.7%) of the respondents were from the low income level. Regarding occupation of the respondents, (49.3%) were students, and (40.4%) of the respondents were employee, while only (8.5%) of them were unemployed.

Table 1. Sample Characteristics: Age Ranges

		Frequency	Percent
Valid	18 - 24	113	41.9
	25 - 31	69	25.6
	32 - 38	40	14.8
	39 - 46	31	11.5
	47 - 55	17	6.3
	Total	270	100.0
Missing	System	0	0.0
Total		270	100.0

Table 2. Sample Characteristics: Accommodation

		Frequency	Percent
Valid	Amiriyah	33	12.2
	Mmansour	32	11.9
	Jihad	32	11.9
	Mamoun	28	10.4
	Doraa	27	10.0
	Saidiyah	24	8.9
	Ghazaliya	23	8.5
	Salehia	23	8.5
	Bayaa	22	8.1
	Utaifiyya	20	7.4
	Total	264	97.7
Missing	System	6	2.2
Total		270	100.0

5.2. Analysis of Criteria of Open Spaces and Parks

Questionnaire format of the green open spaces criteria was a matrix questions regarding respondents' preferences on five criteria of open spaces and parks comprising: design and image, activities and quality in terms of availability of diverse activities with high quality, accessibility and linkage, safety and security, and management and maintenance. The criteria of green open spaces in general have been identified and mentioned through the studies of previous literature in introduction section.

Table 3. Statistic: Criteria of Open Spaces and Parks

		Design & Image	Activities & Quality	Access & Linkage	Safety & Security	Management & Maintenance
N	Valid	268.00	268.00	267.00	269.00	269.00
	Missing	2.00	2.00	3.00	1.00	1.00
Mean		1.99	1.79	1.92	1.90	2.23
Median		2.00	2.00	2.00	2.00	2.00
Mode		2.00	2.00	2.00	2.00	2.00
First Value*		1.00	1.00	1.00	1.00	1.00
Last Value*		5.00	4.00	5.00	4.00	5.00

Values are: 1=Strongly agree, 2=Agree, 3=Undecided, 4=Disagree, 5=Strongly disagree

Table 4. Frequency: Criteria of Open Spaces and Parks

		Design & Image		Activities & Quality		Access & Linkage		Safety & Security		Management & Maintenance	
	Likert	Freq.	Perc.	Freq.	Perc.	Freq.	Perc.	Freq.	Perc.	Freq.	Perc.
Valid	1	75	27.8	106	39.3	81	30.0	79	29.3	57	21.1
	2	147	54.4	125	46.3	136	50.4	139	51.5	136	50.4
	3	21	7.8	21	7.8	36	13.3	47	17.4	34	12.6
	4	18	6.7	16	5.9	12	4.4	4	1.5	40	14.8
	5	7	2.6	0	0.0	2	0.7	0	0.0	2	0.7
	Total	268	99.3	268	99.3	267	98.9	269	99.6	269	99.6
Missing	System	2	0.7	2	0.7	3	1.1	1	0.4	1	0.4
Total		270	100.0	270	100.0	270	100.0	270	100.0	270	100.0

Values are: 1=Strongly agree, 2=Agree, 3=Undecided, 4=Disagree, 5=Strongly disagree

IOP Conf. Series: Materials Science and Engineering **401** (2018) 012029 doi:10.1088/1757-899X/401/1/012029

Mean, median, and mode values of design and image of open spaces are 1.99, 2.00, and 2, where 1 referred to "strongly agree", 2 referred to "agree", 3 referred to "undecided", 4 referred to "disagree", and 5 referred to the value of "strongly disagree" (Table 3). (27.8%) of the respondents strongly agreed that design and image is an important criteria and (54.4%) of the respondents are agree to that, while (6.7%) of the respondents disagreed, and only (2.6%) of the respondents showed strong disagreement. Mean, median, and mode values of availability of high quality diverse activities criteria are 1.79, 2.00, and 2 respectively (Table 3). Where (39.3%) of the respondents strongly agreed that high quality diverse activities is an important criteria and (46.3%) of them agreed to that, while (5.9%) of the respondents disagreed and no one of them showed strong disagreement (Table 4).

On the other hand mean, median, and mode values of accessibility and linkage criteria of open spaces are 1.92, 2.00, and 2 respectively (Table 3). Where (30.0%) of the respondents strongly agreed that accessibility and linkage is an important criteria, (50.4%) of the respondents agreed to that, while (4.4%) of them disagreed, and only (0.7%) of them showed strong disagreement. Mean, median, and mode values of safety and security criteria are 1.90, 2.00, and 2 respectively (Table 3). Where (29.3%) of the respondents strongly agreed that accessibility and linkage is an important criterion, (51.5%) of the respondents agreed to that, while (1.5%) of them disagreed and no one of the respondents showed strong disagreement (Table 4). Meanwhile mean, median, and mode values of management and maintenance criteria are 2.23, 2.00, and 2 respectively (Table 3). Where (21.1%) of the respondents strongly agreed that accessibility and linkage is an important criterion, (50.4%) of them agreed to that, (14.8%) of them disagreed, and only (0.7%) of the respondents strongly disagreed to that (Table 4).

6. Finding Methodology

Findings from the descriptive and frequency analysis emphasize that most influential criterion in open spaces in *Karkh* is activities and quality in terms of the availability of high quality diverse activities, followed by safety and security factor, accessibility and linkage factor, design and image factor, and management and maintenance factor. However, all these standards are clearly influential according to the respondents' answer, which confirm that all these criteria is required in open spaces and parks planning and design in *Karkh* district, Baghdad to increase the number of visits to the proposed space and in turn increase the social interaction. This result is similar to (Holland, Clark, Katz and Peace) results, about the importance of effective entertainment activities, design, maintenance, management, safety and security factors to achieve effective public spaces [12]. The results of the study are also similar to (Skip, Akhir and Omar) findings, which confirmed that good accessibility and linkage, degree of comfort and image, users, and activities and sociability are important factors to determine the successful public park in Malaysia [21].

So achieving the criteria of open spaces and park in *Karkh*, Baghdad is an essential to enhance social interaction there, and that in turn leads to enhance physical, health, mental, social, environmental, and aesthetic aspects for residents and residential area of *Karkhs*' community. On the other hand, this study was conducted as a result of previous studies findings about the lack of open spaces, recreational open spaces and social interaction in Baghdad after the war of 2003 [13-14-19]. Where this study was conducted in an attempt to provide solutions to these issues.

Table 5. Means to Enhance Social Interaction in Green Open Spaces of Baghdad
Source: Authors, 2016

Objectives	Results	Findings
Main Obj. The criteria of green open spaces to achieve social interaction in the city of Baghdad.	*Support Pervious Literatures:* Parks must have an appropriate criteria and components to encourage social interaction.	1. Availability and quality of activities. 2. Safety and security. 3. Accessibility and linkage. 4. Design and image. 5. Management and maintenance.

7. Conclusion

Paper results indicate that one of the means to achieve social interaction in Baghdad nowadays is to provide an open space with an appropriate criteria. So criteria of green open space are an important factor to achieve sound social interaction. These criteria include; activities and quality, safety and security, accessibility and linkage, design and image, and management and maintenance. Where, existing parks in *Karkh* nowadays do not perform their role functionally and effectively this in turn led to decrease the social interaction among residents. Thus, there is a close correlation between the lack of social interaction and the lack of parks or lack of appropriate criteria in the existing parks. The results of this paper are useful reference for urban and landscape planners, architects, social psychologist, and researchers in this fields, so the significance of this paper lies in linking social-psychology to architectural research.

8. References

[1] Al-Hassani 1958 *Ancient and Modern Iraq* 3rd ed (Sidon: Lebanon)

[2] Al-Rajhy C 2006 *Social life in Baghdad from the beginning of the sixteenth century until the fall of Baghdad in 1258 AD* (Makkah: Umm Al-Qura University)

[3] Al-Samarrai Q 2002 The Abbasid Gardens in Baghdad and Samarra 7-12th Century *Foundation of science technology and civilization* 6 2-8

[4] Al-Wardi A 1972 *Social Aspects of Iraqi Modern History; chapter I* 1st ed (Baghdad: National Library of Baghdad)

[5] Al-Wardi A 1974 *Social Aspects of Iraqi Modern History; chapter III* 1st ed (Baghdad: National Library of Baghdad)

[6] Al- Wardi A 1980 *Social Aspects of Iraqi Modern History; chapter IV* 1st ed (Baghdad: National Library of Baghdad)

[7] Al-Wardi A 2007 *Social Aspects of Iraqi Modern History; chapter V* 1st ed (Dubai: House of Paper for publication)

[8] Al-Wardi A 2007 *Social Aspects of Iraqi Modern History; chapter VI* 2nd ed (Dubai: House of Paper for publication)

[9] Creswell J 2013 *Research Design: Qualitative, Quantitative and Mixed Methods Approaches* 4th ed (Los Angeles: SAGE Publications)

[10] Edith D, Joop J and Don A 2008 *International Handbook of Survey Methodology* (Oxford: Routledge)

[11] Fathel A 2015 *Development the area of Baghdad city in the Abbasid era* Alsabaah journal http://www.alsabaah.iq/ArticleShow.aspx?ID=87453 (Visited: 06 October 2017)

[12] Holland C, Clark A, Katz and Peace S 2007 *Social interactions in urban public places* (Bristol: Joseph Rowntree Foundation)

[13] Ihsanoglu E 2007 Assessing the human tragedy in Iraq *International review of the Red Cross* 89(868) 915-927

[14] JAU, IOM and UN 2011 *Urban Baghdad: Impact of conflict on daily life* (Amman: International Organization for Migration) 2-9

[15] Kara B, Tuncay H and Deniz B 2011 Investigating recreational qualities of the parks in Aydın *Procedia-Social and Behavioral Sciences* 19 158-164

[16] Larson L, Whiting J, Green G and Bowker J 2014 Physical activity locations in Georgia: Frequency of use by socio-demographic group *Journal of Outdoor Recreation and Tourism* 5(6) 68-72

[17] Nezlek J, Richardson D, Green L and Jones E 2002 Psychological well-being and day- to-day social interaction among older adults *Journal of Personal Relationships* 2 1350-4126

[18] Poodeh S and Vali A 2014 Investigating the Characteristics of Open Spaces to Enhance Social Interactions in Neighborhood Environments *Journal of Natural and Social Sciences* 3(4) 148-158.

[19] Rikabi N and Ali E 2013 Study of green spaces and its impact on the environment of residential district: Case study: Kathemih, Baghdad *Journal of the planned development* 28(13) 25-46.

[20] Rydgren J and Sofi D 2011 Interethnic relations in Northern Iraq: Brokerage, social capital and the potential for reconciliation *International sociology* 26 25-49.

[21] Sakip S, Akhir N and Omar S 2014 Determinant Factors of Successful Public Parks in Malaysia *Procedia-Social and Behavioral Sciences* 170 422-432.

[22] Uslu A and Gökçe 2010 Social interaction in urban transformation areas and the characteristics of urban outdoor spaces *African Journal of Agricultural Research* 5(20) 2801-2810.

[23] Yamane T 1967 *Statistics: An Introductory Analysis* 2nd ed (New York: Harper and Rao)

[24] Mamaghani N, Asadollahi A and Mortezaei S 2015 Designing for Improving Social Relationship with Interaction Design Approach *Procedia - Social and Behavioral Sciences* 201 377-385.

[25] Matthias and Degenhardt 2015 The effects of urban inhabitants' nearby outdoor recreation on their well-being and their psychological resilience *Journal of Outdoor Recreation and Tourism* 10 55–62.

IOP Conf. Series: Materials Science and Engineering **401** (2018) 012030 doi:10.1088/1757-899X/401/1/012030

The importance of the heritage values of Al-Ukhaidhir palace, Karbala city, Iraq

Omar K Akram[1], Nada F Mohammed Jamil[2], Sumarni Ismail[3], Daniel J Franco[4] and Andreia Graça[5]

[1] PhD Researcher, Faculty of Design and Architecture, Universiti Putra Malaysia, Malaysia
[2] Researcher, Foreign Affairs Office, Kurdistan Region, Iraq
[3] Lecturer, Faculty of Design and Architecture, Universiti Putra Malaysia, Malaysia
[4] PhD Researcher, Faculty of Computer Science and Information Technology, Universiti Putra Malaysia, Malaysia
[5] Master Researcher, Superior School of Technology and Management, Polytechnic Institute of Beja, Portugal

E-mail: sumarni.upm@gmail.com

Abstract. Karbala is distinguished by its rich architectural heritage. Deterioration of historic buildings can affect the city itself as well as the country as general. This research highlights the issue that Al-Ukhaidhir palace, located in Karbala city, is facing. There is no attention from local population, neither experts to protect the architectural heritage values of Al-Ukhaidhir palace. Therefore, there is a dire need for an action to protect the historical palace, aiming to save it for next generations and to increase tourism not just in the city, but for the country in general. After the introduction of the background problem, this study presents a literature review on Al-Ukhaidhir Palace, highlighting their significant heritage values and then summarizing the issues that are presently putting Al-Ukhaidhir Palace into danger. The methodology of this research is based in a case study, which follows a Qualitative method. This study contributes in identification of the significant values of the palace, proposing a theory on the protection of Islamic Architecture Heritage, from the issues that Al Ukhaidhir Palace in Karbala City is facing and highlights the important role of experts and local population in the preservation process of heritage sites with its recommendations. It helps on the protection and documentation of heritage values in Al-Ukhaidhir Palace, having a global impact towards humanity issues.

1. Introduction

Al-Ukhaidhir Palace is considered an ancient defense building with unique architectural values. It is a distinctive feature of military architecture in the Islamic terms of design and engineering, not only for Iraq, but also for the entire Islamic world. It is still an imposing building, however, with large visual damages [7][15]. Al-Ukhaidhir was built in around 750-892AD. The palace represents different civilizations from the past and was built by the ancestor specialists. Its features, in specific its construction materials, are referred by most archaeologists as dating back to Islamic times, especially to the first Abbasid period [9]. A good study driven on the palace was performed by the British Orientalist KA, focusing the existing architectural and historical evidences [18] [22]. The most important of the existing studies about the place was conducted by him, where it was concluded that the palace was built in 161 AH (778 AD), and is considered one of the most important opinions on this subject [24]. According to many scholars

Content from this work may be used under the terms of the Creative Commons Attribution 3.0 licence. Any further distribution of this work must maintain attribution to the author(s) and the title of the work, journal citation and DOI.
Published under licence by IOP Publishing Ltd

who specialize in Islamic architecture and art, the architectural elements of the fortress of Al-Ukhaidhir are fully in line with the development of Islamic architecture on that era [2]. The city of Samarra was built in the year 221 AH (836 AD), the second Abbasid capital built in Iraq after Baghdad. It was found that there is a great similarity between the architecture of the fort and Samarra archaeological buildings [10]. It is confirmed that the fortress of Al-Ukhaidhir was built in Islamic style, through the excavations that took place. The mosque's mihrab in its southern wall was also built. This mosque has the same characteristics as the Islamic mosques [12]. The destruction of Al-Ukhaidhir Palace results in the loss of this wealth forever, being impossible to recover it in the future. It was built in ancient times, representing different ancient civilizations [6]. Its level of destruction comes mainly from the lack of awareness about its important heritage values and it is also possible to identify absent regulation for revival and restoration processes, as a main factor for its ruins [4]. So, the main research question is: What are the significant Heritage Values of AL- Ukhaidhir Palace, Karbala City, Iraq to be protected? Where its objectives are: To identify the significant heritage values of Al-Ukhaidhir palace; To summarize the issues that Al-Ukhaidhir Palace in Karbala City is facing; To highlight the role of experts and local population on protecting Al-Ukhaidhir Palace [14].

2. Literature Review

2.1. Karbala City

The city of Karbala is located southwest of Baghdad city, at a distance of (105) km, and is linked to the province of Baghdad, Babylon, Najaf and Qadisiyah by different roads (29 km) [8]. The location of the city of Karbala is of a great importance to other cities in Iraq, not just because of its similar climatic conditions but also by being located within the most intensive regions [5]. The city of Karbala has emerged greatly after the martyrdom of Hossain PBUH and his companions in the year 61 AH. Karbala consists of three districts and five administrative units between a district center [8] The elimination of Karbala and Husayniyah; Indian district; The district of Ain al-Tamar.

2.2. Al-Ukhaidhir Palace (750-892AD)

Al-Ukhaidhir Palace ruins geographically located in western sahara after the 48km to the south – west of the city of Karbala, 150 km south of Baghdad and 20 km from the town of Ain Al Tammer called Statha. It was an important stop on regional trade routes [1]. Historically, it possible to find relevant architectural styles and features in Mesopotamia's ancient time [16]. The world has witnessed for thousands of years, even before the Iraqi construction activity had taken place, that creativity and construction, despite the lack of material, had already drawn the attention of different scholars. Being clear that it deserves all admiration and appreciation. From historic figures, Al-Ukhaidhir Palace, which will be highlighted, is considered a unique ancient defense building [17].

Al-Ukhaidhir is a large, rectangular fortress with a unique defensive style (Figures 1 and 2). It consists of a main hall, large Iwan, reception hall and servants quarter [19]. The Al-Ukhaidhir was regarded as an important workshop, where many elements, of what was to become known as Muslim architecture, were elaborated and developed [22].

2.3. The Significant Heritage Values of the Palace

The architects of Al-Ukhaidhir first introduced a new elaborate technique based on the construction of elliptical (pointed) barrel vaults, with bricks in similar technique to building a wall and, therefore, considerably eased the way vaults were built [24]. The old tradition of building, using mixture of mortar and small stones and debris, laid out on wooden base. Such method need a lot of wood, not available in this arid region, where building took considerable time to be finish, as workers had to wait for the vault to dry, to move the scaffolding form

Figure 1. Pointed barrel vault and arches of the great hall of Al-Ukhaidhir Palace – Source: https://www.cemml.colostate.edu

Figure 2. The Northern side of the court of honor and the upper floor of Al-Ukhaidhir Palace – Source: https://www.cemml.colostate.edu

to another part of the building. This new technique, likely to have been introduced through Persian and Mesopotamian works and methods, provided adequate solutions to these issues [23]. Another elaboration of the vault construction technique was seen in the palace's mosque, the use of flattened arches, to upholding the brick vault, was a technique that became later known as the ribbed vaulting. According to Marcais, this method was also used in Medinat Al-Zahra (10th century) in Andalusia. So, such methods, later provided the foundation for the rise of Gothic architecture in Europe [25]. Blind arcades are also found in Al-Ukhaidhir place, being them another original element introduced in Al-Ukhaidhir as a decorative form in the Northern façade of the Honor Court. This feature became an essential element in Muslim architectural decoration and later this method was transmitted to Europe [24]. Here, one has to point to the attitude of Western scholars which connected the use of blind arcading, as well as the origin of ribbed vaulting and a number of other features, to Lambard architects, who are considered (by them) to be the main builders of Dark ages Europe, especially in the 10th and 11th centuries [21]. Fluted dome was another innovation that appeared crossing beyond the main entrance in Al-Ukhaidhir Palace [2]. Al-Ukhaidhir elaborated the defensive technique by introducing what's known as "chemin de ronde" along the ramparts. Another elaboration of the four gates, consisting each of a chamber with an inner wall and an outer portcullis, which could be lowered in case of assault, trapping the attackers inside, provided another defensive architectural technique that was also transmitted to Europe through the crusaders [19]. The palace consists of two fortified enclosures. On the outer enclosure, there is a 17 meters high robust rampart made of limestone slabs and mortar strengthened by corner and intermediary towers, alternating with pairs of blind arches on pilasters [14]. A series of portcullises, or cataracta, are usually suspended by iron rings to be dropped behind the enemy, imprisoning him, were fitted along it and even on the supporting towers, adding extra protection [11]. These defensive arrangements have added extra protection. Another defensive scheme employed in Al-Ukhaidhir was the use of arrow-slits, which were served by the wall walk (once vaulted). From these slits defenders of the palace it was possible to fire arrows without being exposed to the enemy. The gateways were also fitted with slits in their vaults for dropping flaming missiles, as well as lateral grooves for letting down the portcullis [21].

2.3.1. Internal Facilities The facilities are confined to a rectangular building adjacent to the northern wall, 112 meters long and 80 meters wide and has the following internal facilities:

The official Diwan (large lobby):
It is a rectangular hall shape length 15.50 meters and width of 9 meters. In its eastern and western sides are four semi-cylindrical arches. The entrance on the eastern side leads to the upper floor, while the three entrances lead to rooms. In the last Iwan there are two entrances, the first leading to the upper floor, the second leading to the mosque and the third consisting on a group of overlapping rooms [17].

Central Section:
This section surrounds the hall of the ceiling, separated from the rest of the palace. Accessed through an entrance in the middle of the southern side of the official Diwan (Large lobby), offset by another entrance that leads to a grand spacious. It is a rectangular, 33-meter-long, 27-meter-wide courtyard with four doors open by the grand gallery. The northern entrance is connected to the Great Iwan. The eastern and western entrances lead to the surrounding area of this section, including the Haram section. The southern entrance is directly connected to the so-called "mabin", and the northern rib of the sediment is higher than the rest of the ribs. The eastern and western sides consist on cavities of a single pattern. The southern side consists on a large opening, which is a facade of the large Iwan [20].

The Large Iwan:
It is a rectangular hall with a length of 10.75 meters, and 6 meters wide, topped by a semi-cylindrical built of brick, surrounded by a number of halls for the reception of guests [15]. The façade presents a beauty design, as if it was designed in the style of a throne room. The Iwan has two halls on its eastern side and two others on its western side. These halls are rectangular, 10 meters long and 4.65 meters wide each. They are characterized by a series of various castles and decorations. The Iwan has also a rectangular courtyard that is surrounded by a number of halls. In the middle of the southern side of this courtyard, there is an entrance leading to the large gallery that surrounds the central section and opens in this gallery several doors [16].

The Mosque:
It is a rectangular building, 24 meters long, 1550 meters wide, with an entrance in the lobby and another in the western facade, which separates the fort wall from the northern section of the palace and its facilities. Surrounding the mosque from its sides, the front, eastern and western corridors, it is based on a row of columns in each hand, on which the arches are consecutive and complexly decorated with beautiful geometric forms. The shape of the mihrab is concave [7]. Archaeological investigations have shown that its construction is an integral part of the construction of the Al-Ukhaidhir fort and not modern or added to in a later time [22]. The mosque has three entrances: two of them are led by the northern passage and are located on the northern side of the mosque's courtyard. The third entrance is located in the south-eastern corner of the mosque and leads to the lobby [24].

Guard Section:
Located on the east part of the lobby, and adhered to the outer wall to the north, it leads to an entrance located in the long passage [17]. There is also another entrance in the form of a sloping surface leading to the first floor, by the entrance to the branch on the right, and to the second floor after turning south, up to the second floor. This section consists of one floor, except the western side, which consists of two floors to connect rooms and facilities for the guards. It is located at the top of the main entrance [23].

2.3.2. Residential Houses They are four houses that are separated from one another. Two are located to the east of the central section. The other two are located to the west of this section. In each house, there is an entrance located in the grand gallery. These houses are almost identical in terms of design, symmetry and location [19]. Each house consists on a rectangular dish, and its sides are slender. These arches are topped with arches based on columns and the entrance of the house is in the central courtyard. The northern and southern facades are often similar in design, number of halls and facilities. These houses are characterized by the Awawin system, a model attributed to the architectural style of Al-Heiri (relative to the city of Al-Hira) near the city of Kufa [8].

2.3.3. Servant Section It is a large building consisting on a rectangle, with eight rooms with vaulted ceilings, isolated from the family houses and located behind the central section and the reception rooms [21].

2.3.4. The Bathroom A rectangular building is located between the south house of residential houses and between the servants' section. It is 9.90 meters long and 9.30 meters wide and its entrance is located in the south-eastern corner of the Grand Gallery [19]. The bathroom is surrounded by an internal wall and, at the same time, by the main wall of the palace. This bathroom is special for the palace, where its interior walls are built of brick and stone and its floor is tiled with white marble. It has an internal main entrance located in the south-eastern corner of the Grand Corridor apart from other available ones. The shower section consists of three interconnected rooms, leading to an entrance in the western side. The overall design of the bathroom is similar to the ones built in the Islamic ages [22].

2.4. External Buildings
There are buildings outside the fort. The first is located on the right of the interior north (main) entrance. It is rectangular with 76 meters long and 12 meters wide, consisting of a series of 14 rooms with vaulted ceilings. A door to a ladder leads to the roof of the building [15]. There is a 33.20-meter-long rectangular courtyard with a width of 24.70 meters between the rectangular building on the northern side and the wall on the western side [17]. There is also a small building on the north-eastern side of the fort, similar to the design of the bathroom [19].

2.5. Proposition Theory
Protection of Islamic Architectural heritage in Al-Ukhaidhir palace, increases the impact of heritage tourism to the city, gives opportunity to improve the country's economic values, while keeping it to next generations [14].

3. Methodology
The qualitative method is effective on this research in the way that it is used to obtain information about Al-Ukhaidhir, like the historical background about the palace, such as significant architectural heritage, internal facilities, among others. In order to have better understanding on the needs of protecting the significant heritage palace the exploratory case study adopts the most common qualitative approaches, based on main research question of this research and the study's nature. This approach aims to identify their problems and to find sensitize experts and local population for the need to protect them from possible loss [14].

4. The Main Problems Facing to Protect the Historic Place
The preservation of the architecture heritage faces a large number of problems that can be attributed to social and economic development [2] Lack of awareness of the importance of urban heritage; lack of funds for conservation; absence of legal protection; absence of community

and private sector participation and failure on the development of appropriate mechanisms to increase community and private sectors' participation in the conservation process.

5. The Role of Experts and Local Population in the Heritage Sites' Preservation Process

The involvement of the local community and the expert group have the greatest impact on the support, development and revitalization of Al-Ukhaidhir Palace. Therefore, it is necessary to document and create clear policies for communication between the local community and authorities in order to develop marketing plans, whether at the local or global level, and to provide services that can help on attracting tourism in its various aspects [3][?]. The country should seek to develop laws on investment, protection, restoration and revitalization, in order to encourage the local community to participate and not only to give them possibilities to be part of preservations processes, but also to give them opportunities to have jobs and to prevent crime and vandalisms in that place [12, 14]. The preservation must work under control of authorities and specialists, and the local population should follow the regulations of government to be unified in all state. With this, the preservations process may be fulfilled [15].

6. Conclusions and Contributions to Existing Knowledge

- It is recommended to prevent the demolition or alteration of archaeological heritage areas except through clear studies and plans that take into account the values, importance and optimal methods of dealing with them;
- It is recommended to prevent infringements and structural irregularities and random elements that distort areas and heritage palace;
- The importance of media and awareness to the local populations about heritage sites.
- The conservation management projects in a manner that ensures continuity and maintenance after the completion of conservation work;
- The development of laws and regulations governing the treatment of the palace;
- Predicting all causes of deterioration and taking appropriate measures to stop them in a timely manner;
- Determine the levels of impact and treatment methods in order to preserve the architecture heritage for Al-Ukhaidhir Palace.

Based on the three different objectives, the aim of this research is to identify, summarize and highlight the role of experts and local population on protecting Al-Ukhaidhir Palace, located in Karbala City, Iraq, in a way to help their protection and provide reliable information for future generations and future studies on the area. Documenting the significant heritage values in Al-Ukhaidhir is a vital factor for its identification, protection and interpretation:

- Identification: identification of the significant heritage values of Al-Ukhaidhir palace, provided good and reliable documentation for future generations and studies;
- Protection: The lack of security on the country and a possible unexpected attack by other terrorist groups, can ruin all heritage values in palace and its Islamic Architecture styles, which made it urgent to properly document them;
- Interpretation: Documenting the significant heritage values of Al-Ukhaidhir palace, provided an overview and assessment of the current state of the physical and cultural aspects of the properties.

The obtained results, namely about the significant values of Al-Ukhaidhir palace, deliver important information for students, researchers, planners and even to visitors, providing them a better knowledge and understanding of the place and the importance of its protection. This

IOP Conf. Series: Materials Science and Engineering **401** (2018) 012030 doi:10.1088/1757-899X/401/1/012030

study may be used on foundation to conservation, rehabilitation and reconstruction of the heritage present in Al-Ukhaidhir palace.

References

[1] Lima M G 2011 Islamic art and architecture an introduction to the architecture and the arts of the civilizations of islam
[2] Akram, O K 2017 Identification of cultural properties of arab district, erbil, iraq *University Putra Malaysia*
[3] Akram O K, Franco D and Ismail S 2016 Smart buildings – a bew environment (theoretical approach) *International Journal of ENgineering Technology, Management and Applied Sciences* **4(4)** 1–5
[4] Akram O K, Franco D J and Ismail S 2016 The cultural significant of erbil city: case of traditional kurdistan houses *International Journal of Engineering Technology, Management and Applied Sciences* **4(2)** 102–106
[5] Akram O K, Franco D J, Ismail S, Muhammed A and Graça A. 2016 Promoting heritage management in new smart cities: évora city, portugal as a case study *International Journal of Engineering Technology, Management and Applied Sciences* **4(9)** 148–155
[6] Akram O K, Ismail S and Franco D J 2016 The heritage values of arab district, erbil city, iraq *International Journal of Engineering Technology, Management and Applied Sciences* **4(2)** 129–135
[7] Akram O K, Ismail S and Franco D J 2016 The significant of islamic architecture heritage at baghdad city, iraq – case studies of shrine of lady zumurrud khatun *International Journal of Engineering Technology, Management and Applied Sciences* **4(4)** 133–138
[8] Akram O K, Ismail S and Franco D J 2016 The significant of tourism heritage of najaf city in iraq *International Journal of Engineering Technology, Management and Applied Sciences* **4(4)** 14–19
[9] Akram O K, Ismail S and Franco D J 2016 The significant of urban form of erbil city, iraq *International Journal of Engineering Technology, Management and Applied Sciences* **4(2**, 96–101
[10] Akram O K, Ismail S and Franco D J 2016 Urban and tourism development projects for cities citadels (aleppo and erbil) *International Journal of Engineering Technology, Management and Applied Sciences* **4(3)** 60–68
[11] Akram O K, Ismail S, Utaberta N, Mohd Yunos M Y, Ismail N A and Arifin N F 2015 The important values of architectural baghdadi heritage *Advances in Environment Biology* **9(24)** 46–52
[12] Franco D J, Akram O K, Ismail S, MohammedJamil N F and Graça A 2017 Adopting the smart buildings system to heritage buildings In melaka, malaysia *11th ASEAN Postgraduate Seminar* 01–09
[13] Akram O K, Franco D J and Ismail S 2016 Development phases from heritage buildings to smart buildings *International Journal of Engineering Technology, Management and Applied Sciences* **4(4)** 6–13
[14] Akram O K, MohammedJamil N F, Franco D J, Graça A and Ismail S 2018 How to guide your research using ondas framework
[15] Creswell K A 1958 *A short account of early muslim architecture* (Middlesex Baltimore: Penguin Books)
[16] Hattstein and Delius 2000 *Islam Art and Architecture* (Cologne, Konemann)
[17] Reimer D, Vohsen E 1923 *Forschungenzur islamischen kunst* (Berlin)
[18] Jairazbhoy R A 1972 *An outline of islamic architecture* (Bombay and London, Asia Publishing House)
[19]] Kritzeck J 1959 *The world of islam: studies in honour of Philip K. Hitti* (London, Macmillan)
[20] Lethaby W R 1904 *Medieval art: from the peace of the church to the eve of the renaissance* (New York, Duckworth and Co London) IV 100-111
[21] Marçais G 1954 *L'architecturemusulmaned'occident: tunisie, algérie, maroc, espagne et sicily* (Paris, Arts et métiers graphiques)
[22] Porter K A 1909 *Medieval architecture, its origin and development* (New York, Baker and Taylor Company) 1
[23] Rice D T 1979 *Islamic art* (Norwich, Thames and Hudson)
[24] Richmond E T 1926 *Moslem architecture 623-1516 some causes and consequences* (London, The Royal Asiatic Society)
[25] Rivoira G T 1918 *Moslem architecture: its origin and development* (G. Oxford University Press)

IOP Conf. Series: Materials Science and Engineering **401** (2018) 012031 doi:10.1088/1757-899X/401/1/012031

Neighbourhood Safety and Outdoor Play Activities among Urban Children in Shah Alam, Malaysia

Habizah Sheikh Ilmi[1], Mariatul Liza Meor Gheda[2] and Nooridayu Ahmad Yusof[3]

[1]School of Architecture, Building and Design, Taylor's University, Malaysia

[2]Faculty of Design and Architecture, Universiti Putra Malaysia, Malaysia

[3]Wels School of Architecture, Cardiff University, United Kingdom

Email: Habizah.ilmi@taylors.edu.my

Abstract. Most of the generation born in the 70s to the 80s usually are attached to their memories of bathing and playing in the river, climbing trees, playing slingshot and catching fish in the pond without an adult supervision. Children nowadays do not have the privilege of experiencing the same thing. Today, they no longer have the freedom to roam around and enjoy the outdoors, as there is a high exposure to danger that comes with the development. The aim of this research is to understand the connection of neighbourhood safety with outdoor play activities among urban children. Data were derived from focus group interviews and children's drawings that were gained from a research participatory workshop in a primary school and a day care center in Shah Alam. Children aged 10 and 11 from the school were selected as the respondents in this research. This research found that a stranger harm came out as the most significant issue that affect children's outdoor play in urban neighbourhood. In Shah Alam, neighbourhood safety seems to be related to outdoor play activities. Outdoor play activities may be best encouraged by improving different safety components, rather than focusing one safety aspect such as traffic safety.

1. Introduction

Children development is bound to face plenty of problems in the city, particularly in industrialized and relatively developed cities (Gökmen and Taşçi, 2016). The outdoor environment for many children is very important as their place to play and learn and many research show a marked preference amongst most children for outdoor play. It could be argued that among primary school children, active free play or unstructured physical activity that takes place outdoors in the child's free time may potentially be the major contributor to children's physical activity (Burdette et al., 2004).

Janssen and LeBlanc (2010) determined that a physical activity during childhood is important for psychological benefits. The concept of Child Friendly City shall encourage local governments to make decisions that are in the best interests of children and promote children's rights to a healthy, caring, protective, educative, stimulating, non-discriminatory, inclusive and culturally rich environment (Malone, 2001; Riggio, 2002).

Moreover, the relationship between the neighborhood and outdoor home physical environment and early child health and development may be mediated by 'behaviors' that facilitate early child health and development such as play, physical activity, social interaction, exploration and stimulation. Reviews of the correlates of children's physical activity and outdoor play indicate that built environment features such as walk/bicycle paths, presence of cul-de-sac roads, access to parks, recreational facilities, other local destinations and public transport are positively associated with children's physical activity, while high traffic exposure and crime are negatively associated with children's physical activity (Davison and Lawson, 2006). Nevertheless, they provide as a guidance to the pathways through which the built environment influences child development. For example, neighborhoods characterized by high street traffic and a lack of local attractive parks may restrict children's opportunities for play and interaction with other children in common play spaces such as the front yard and local park. This leads to where the children are imprisoned at their houses or their

Content from this work may be used under the terms of the Creative Commons Attribution 3.0 licence. Any further distribution of this work must maintain attribution to the author(s) and the title of the work, journal citation and DOI.
Published under licence by IOP Publishing Ltd

friend's houses and playgrounds offered by commercial spaces (Mc. Kendrick, cited in Gökmen and Taşçi, 2016).

2. Research Method

In order to understand children's perceptions interactive data collection is considered beneficial to gain real life responses. For this study, the research approach has been adopted in the form of a case study in Section 7, Shah Alam, Malaysia. This approach was selected for this research as it focused on a unique case that requires the collection and analysis of qualitative data. The justification of this adoption lies in the fact that the research needs the value of depth over quantity. This research approach works at delving into social complexities in order to truly explore as well as understand the interactions, processes, lived experiences and belief systems that are a part of individuals, institutions, cultural groups and even the everyday (O'Leary, 2010). The research data gathered via focus group interviews and drawings of 40 children aged 10 and 11 from a research participatory workshop in SK Seksyen 7 and Transit NurKasih, Shah Alam.

Table 1. The List of Participants

Age	Sekolah Kebangsaan Seksyen 7, Shah Alam		Transit Nurkasih, Seksyen 3, Shah Alam	
	Boy	Girl	Boy	Girl
10	7	7	3	3
11	7	7	3	3
Total	40			

Table 2. The Code Number of Participants

Code No	Gender	Age	Code No	Gender	Age
BA1	Boy	10	GA1	Girl	10
BA2	Boy	10	GA2	Girl	10
BA3	Boy	10	GA3	Girl	10
BA4	Boy	10	GA4	Girl	10
BA5	Boy	10	GA5	Girl	10
BA6	Boy	10	GA6	Girl	10
BA7	Boy	10	GA7	Girl	10
BA8	Boy	10	GA8	Girl	10
BA9	Boy	10	GA9	Girl	10
BA10	Boy	10	GA10	Girl	10
BB1	Boy	11	GB1	Girl	11
BB2	Boy	11	GB2	Girl	11
BB3	Boy	11	GB3	Girl	11
BB4	Boy	11	GB4	Girl	11
BB5	Boy	11	GB5	Girl	11
BB6	Boy	11	GB6	Girl	11
BB7	Boy	11	GB7	Girl	11
BB8	Boy	11	GB8	Girl	11
BB9	Boy	11	GB9	Girl	11
BB10	Boy	11	GB10	Girl	11

For children under the age of 11, visual technique has been proven to be very helpful in the questioning process. This is because of pictures make the subject much clearer than the verbal representation alone (Robinson, 1986). Discussion with individual children about their drawings and notes on the drawings were recorded. To understand how children felt about their neighbourhood and the outdoor play activities, the focus group interviews were conducted. In this activity, participants were required to draw the map of their existing neighbourhood based on what are provided around their houses. These include the things that they like, as well as what they dislike. Regardless of what facilities have been provided within the neighbourhood, the participants had to try their very best to express them in the maps.

The participants could always ask the facilitators if there were anything that they were not sure of. As all these children have different styles of drawings, some of the drawings were neat and can be easily understood, however, some of the drawings were difficult to be deciphered. However, regardless of the outcome, each of the drawings has been included as part of the data obtained in this study.

Having done that, at the end of the session, together with the facilitators, the author interviewed each and every participant, in order to interpret the maps drawn by them, as well as to ensure that there were consistency in what was understood with what the participants intended to express.

In the focus group interview, the questions were highlighted on how the children feel about their neighbourhood safety and outdoors play activities; who they usually play with; what make them enjoy and be happy; what they are scared of, and where favourite places are for them to play and how good their neighbourhood safety condition is.

3. Results

A good street design is central to socially well-functioning neighbourhoods. The prime factors influencing sociability at the street level are scale and safety. Scale is in the size of buildings and their relationship to the street and safety as in pedestrian safety (Freeman & Tranter, 2011). This echoes the study done by Hart (2008), through which he discovered that residents living on heavy traffic streets tend to have 75% fewer friends on the street than those living on similar streets with less traffic. Families on heavy traffic street limit their children's time outdoors and tend to have fewer social connections.

The focus group discussion were revolved around their feelings about living in their neighbourhood, through which the participants have expressed their likes and dislikes, as well as the reasons for those. For example, **BA3** have expressed his dislike for his neighbourhood due to constantly feeling unsafe when playing outside his home, even in front of his own house. His fear was mainly attributed to the fact that his house was located near the main road, with constant traffic moving in and out of the area. He felt that it exposed him to the danger of being hit while playing.

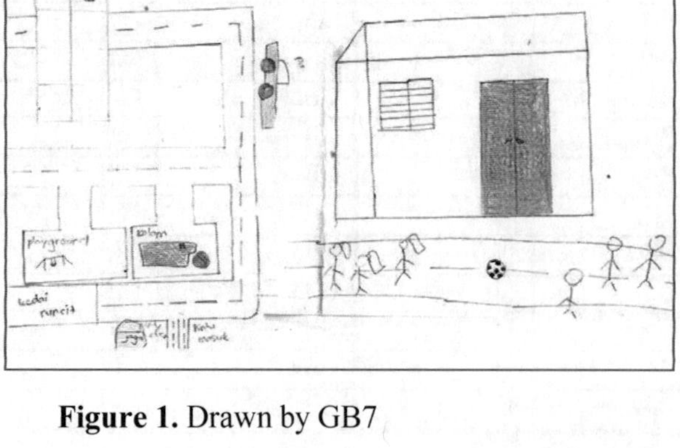

Figure 1. Drawn by GB7

GB7 lived in the fifth floor of the Kristal Apartment. From what has been portrayed, though there were a playground and a swimming pool located within the area, she still felt lonely because she had no friend. As seen in one corner of the map, there were three boys playing a ball game and three girls playing in an area. When asked, she explained that the children belonged to the area of the link houses, just opposite where she was living. She often saw other children playing on the street in front of their house. But, she never plays down there because it is too dangerous and parents did not allow her to go.

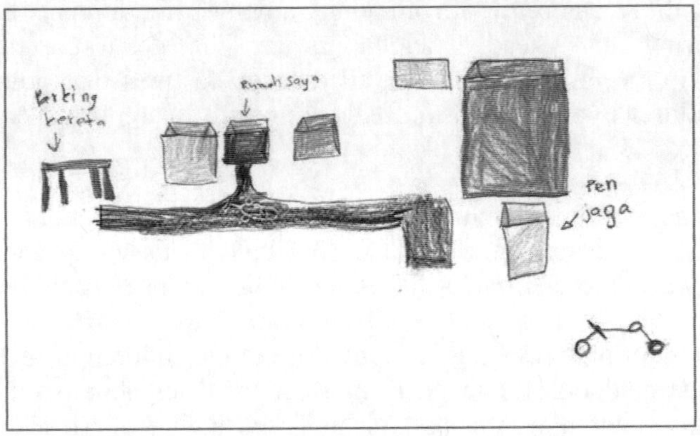

Figure 2. The neighbourhood map as drawn by BB5

However, **BB5** thought that he was lucky because he lived in a government's quarters. The security guard is provided in the quarters (see Figure 2). He can freely cycle within his neighbourhood during weekends.

I love outdoor play because the space is wide, has fresh air and what more exciting is that I can cycle too. Normally, I will spend time outdoor during weekend evenings. (Translated by the author)

When looking into the children's perspective, the author discovered that children wish to play freely without the presence of strangers. They have been taught to always be suspicious to anybody that they have never engaged with. These strangers are not specific to some children, but they include all men, women, teenagers, adults, old people and even those at their age. What they know is that they cannot simply trust anybody except their parents, siblings and family members. From the focus group interview, it can be seen that parents nowadays worry about the environment because of the increased crime rates. No doubt that sometimes, the publicity of these dangers is often exaggerated. Nonetheless, parents are anxious and do not want to lose their beloved children. As mentioned by **GA9**:

Well, mama let me out but not too far from home because the bad people may be outside there. Fortunately, I don't really like going out alone unless for outdoor play with friends. (Translated by the author)

Satisfaction related to the built environment is often linked with perceptions of the neighborhood (McCulloch, 2012) including safety (Buys & Miller, 2012). Neighborhood safety concerns may influence family practices and parental restrictions, thereby affecting children's opportunities to play outside (Molnar et.al 2004) and interact with others locally (Carver and Timperio, 2008). When talked about these strangers, **BA8** mentioned that:

I'm staying in the flats... The playground and the field are just nearby...But. I feel much safer playing indoor because I think my neighbourhood is not so safe. I heard about burglary cases in my neighbourhood. I'm scared of these bad people. (Translated by the author)

IOP Conf. Series: Materials Science and Engineering **401** (2018) 012031 doi:10.1088/1757-899X/401/1/012031

The girls on the other hand preferred to have their parents to accompany them, as they feel safer that way. The issue of stranger harm seems to be a concern here even though the evidence illustrated that there are little likelihood for stranger danger encounters. This may just be a residue of a 'cultural' fear of strangers pervading Malaysian media.

Passive surveillance is also reduced by low sense and connection with the community. Thus, a strong sense of community among the residents could increase the sense of belonging amongst the community and to the place itself. Since most of the parents here may be living far from their relatives, therefore, the strong sense of community amongst the residents is very important. Having a loving family, caring and friendly neighbours are the greatest support in a child friendly neighbourhood environment. On a more detailed level, features that promote a safe and nurturing environment for children to play and move freely are positive parents and community support.

4. Conclusion

Based on the findings, it can be concluded that the opportunities for children in urban neighbourhoods to join in safe play are decreasing. The data from both focus group interviews and the children's drawing from the workshop illustrated that issues about stranger harm are more than the issues of traffic danger with children in Section 7, Shah Alam. This has affected the children's social life, particularly their outdoor play activities. The majority of the children have expressed that they want to be with their friends outdoors. However, in most of their neighbourhood, the distance between the places of interest for children such as park, school, and homes of friends is generally very big. It is recommended that further research be undertaken to study the type of housing setting that best suit for children that they would enjoy the outdoor play activities.

References

[1] Burdette HL, W. R 2004 Parental report of outdoor playtime as a measure of physical activity in preschool-aged children. *National Center for Biotechnology Information, U.S. National Library of Medicine.*

[2] Buys L & Miller E 2012 Residential satisfaction in inner urban higher-density Brisbane, Australia: role of dwelling design, neighbourhood and neighbours. *Journal of Environmental Planning and Management*, Vol 55 (3), Taylors and Francis.

[3] Carver A, Timperio A, and Crawford, D 2008 Playing it safe: The influence of neighbourhood safety on children's physical activity. A review. *Health & Place*, 14 (2), 217-227.

[4] Davison KK1 and Lawson C 2006 Do attributes in the physical environment influence children's physical activity? A review of the literature. *US National Library of Medicine National Institutes of Health.*

[5] Freeman C and Tranter P 2011 *Children and their urban environment: changing worlds*. London: Earthscan.

[6] Gökmen H and Taçşi B G 2016 Views about Child Friendly City: A Case Study from Izmir. *Megaron*. 2016, Vol. 11 Issue 4, p469-482. 14p.

[7] Hart J 2008 Driven to Excess: Impacts of Motor Vehicle Traffic on Residential Quality of Life in Bristol, UK. School of Build and Natural Environment. Bristol: University of the West of England.

[8] Janssen L and Leblanc A 2010 Systematic review of the health benefits of physical activity and fitness in school-aged chidlren and youth. *International Journal of Behavioral Nutrition and Physical Activity.*

[9] Malone K 2001 Children, youth and sustainable cities. *Local Environment*, 6 (1), 5-12.

[10] McCulloch A 2012 *Housing density as a predictor of neighbourhood satisfaction among families with young children in urban England. Population, Space and Place*, Vol. 18 (1). Wiley.

[11] Molnar B E, Gortmaker S L, Bull F C, et al. 2004 Unsafe to Play? Neighbourhood disorder and

lack of safety predict reduced physical activity among urban children and adolescents. *American Journal of Health Promotion*, 18.
[12] O'Leary Z 2010 *The Essential Guide to Doing Your Research Project*. London: SAGE Publications Ltd.
[13] Riggio E 2002 Child friendly cities: good governance in the best interests of the child. *Environment and Urbanization*
[14] Robinson W 1986 *Children's Understanding of the Distinction between Messages and Meaning*: *Emergence and Implications*. Cambridge.

An analysis of women's access and participation in the mosques in the contemporary world

Nangkula Utaberta[1*], Fariba Samia Omi[1], Siti Balkish Roslan[2], Noor Fazamimah Mohd Ariffin[3]

[1] Department of Architecture, Faculty of Design and Architecture, Universiti Putra Malaysia

[2] Department of Architecture, Taylors University, Malaysia

[3] Department of Landscape Architecture, Faculty of Design and Architecture, Universiti Putra Malaysia

Corresponding Author: Nangkula Utaberta
Email: nangkulautaberta@gmail.com

Abstract. In Muslim countries, mosques that designate areas for women's prayer rarely allow women to line up directly behind men in the same hall, as was the practice in the Prophet's Mosque during his lifetime. In many Arab countries, it is common to separate men's prayer place from women's, particularly in smaller mosques. The distance between the two prayer areas vary from one mosque to another. The majority of mosques in Muslim countries house specified places of prayer for women, which take shape of small rooms in the basement, on the ground floor, in a closed balcony or in a small building attached to the mosque. This paper aims to analyze the accessibility, permeability and scope of participation of women in the Mosque in the current scenario of modern world.

1. Introduction

The mosque is a place where Muslims can come together for religious practices as well as it also serves as a social space, allowing the formation of an individual, social interactions and collective memories. Its spatial structure and location in any urban or rural area is formed to cultivate enduring practices, organize daily life and increase social interactions among different socio- economic groups. Earlier, people used the mosque not only for prayer but also for many other activities such as the mosque functioning as a 'school' where the community learned their religion or as a 'parliament' where the society discussed new laws and affairs of the state. It also served as courthouse where judgments were passed and as community centers where neighbors are able to meet with one another, share their problems and hold their celebrations. In short, it was the hub of public life for the emerging Muslim nation.

Content from this work may be used under the terms of the Creative Commons Attribution 3.0 licence. Any further distribution of this work must maintain attribution to the author(s) and the title of the work, journal citation and DOI.
Published under licence by IOP Publishing Ltd

IOP Conf. Series: Materials Science and Engineering **401** (2018) 012032 doi:10.1088/1757-899X/401/1/012032

Regarding this issue, the present crisis is the seclusion of the female group of the community from the premise of the mosque. This unfortunate phenomenon did not prevail during the time of early Muslim generations. Reported by (Hamid, 1973), during the time of the Prophet (pbuh) and the earlier generation of Muslims, women attended mosques, as did men for all kinds of prayers and activities, but with the development of orthodox notions, women's participation in the mosque were being questioned by the teachers of morality.

2. Problem

The mosque is a place for spiritual growth and development for all Muslims, and should be equally accessible for both genders, but the present-day scenario is far from the ideal picture. According to the opinion of (Hamid, 1973) children, apart from receiving religious education, can even play inside the mosque. As Imam Ghazzali wrote, "For children, there is no harm if they enter the mosque and it is not unlawful for them to play inside the mosque. To remain silent at this scene is not unlawful provided that the mosque is not turned into a playground."

The same welcoming environment should be maintained for the women. As Spahic Omer said, the first generation of Muslims capitalized every opportunity to acquire and disseminate knowledge. Some of them were so devoted to it that they were ready to sacrifice some things of which they cherish. That is why the Prophet's (pbuh) mosque was always bustling with life. Study circles, intellectual discourses, meditation etc. made the mosque virtually never devoid of people. Neither women nor children have been disregarded in this process. Even the Prophet (pbuh) designated time of the week to exclusively teach the women and listen to their own issues in a comfortable manner secure from the intrusion of men. (Omer, 2013)

The statement illustrates the importance of incorporating women in the mosque as well as the children of the society to achieve the holistic manner of serving as a center of community development.

Unfortunately, the present situation is different to the ideal one. Reported by Svetlana Peshkova in an interview of an Uzbek woman who said, "We have a desire to go to the mosques, but in Uzbekistan only men go there. We [women] are not very upset because the Qur'an says that one needs to read within a group and we do just that at our meetings." (Peshkova, 2009). Similar situations have been reported regarding other countries as well where women are not allowed to enter into the mosque or participate in the activities therein. (Nawaz, 2005)

Women were allowed to pray along with men during the times of Prophet Muhammad (pbuh). Even though, it was a rule for men to go to mosques to pray five times, he was lenient with women stating that their prayers are as good as the ones in mosques even if performed from the home. Women were active in public life and regularly attend the prayer at the mosque including *Fajr* and *'Isha*, during the time of Prophet Muhammad (pbuh). Children and babies accompanied their mothers to the mosque. The Prophet (pbuh) explicitly stated that men should not prevent the female slaves of Allah from entering the mosque of Allah, even at night. At the same time, the Prophet (pbuh) did not make it a rule that women must pray in the mosque. (Reda, 2015)

Given the Prophet's (pbuh) advice and example there should never be a mosque that tells women to leave when they want to enter and participate in prayers or other activities. It is a woman's right to choose whether she wishes to participate in mosque activities, and she must have open access to the mosque. In accordance to the argument above, Islam does not forbid women from entering mosques or praying along with men shoulder to shoulder. However, it has become the norm in placing rules on women as such. This research will discuss the presented issue, the understanding of a women's role during the Prophet's (pbuh) time and how the space of prayer for women developed.

The Quran stresses that both men and women have a moral obligation to develop themselves spiritually and morally and to fulfill their social responsibilities. The mosque is, and has always been, the center of moral and spiritual learning and growth. It is a public place for discussing issues of public concern and to respond to challenges facing the community. The Quran is also clear on the equal responsibilities of both men and women for developing the public good but today the physical space and facilities available to women in our mosques are not always sufficient. Women are an extensive and imperative part of the population and can play an important role in the development of the community by participating in social activities. Activities can start from the mosque where women participate in social, educational, political, religious and economical purposes, as development can be rather more constructive with women's contribution in the society. Mosque should act as a primary area of where they can gather, interact and take part in social development along with prayer.

The present scenario does not mirror the ideal role of a women and her relationship with the mosque. Misconceptions on the entering of the mosque and attending activities within the mosque, whether in relation to prayer or other exercises, are scrutinized. The patriarchal driven mass have preferred women to perform prayers within the compounds of the home, cutting access to knowledge and cultivating ignorance. Adding to the problem are the proposed and built designs of current designs of mosques, which do not consider women's needs in the application of spaces. As a result, the allocated space of prayer and activity become desolate.

However, the majority of mosques in Muslim dominant countries do allocate specified halls or small rooms for women on the ground floor, in an enclosed balcony or in an extended space of the building, though seeming like an afterthought. Speakers are used to relay and communicate the Imam's voice when prayers and sermons are in session. The halls are smaller in comparison to the main prayer hall, are not properly equipped with basic facilities, poorly designed and sacrifices comfort and connectivity with the main prayer hall. As a result, the women naturally feel less welcomed and privileged in comparison to their counterpart. In other scenarios, the women are discouraged to enter from the main gate and alternatively forced to take a lesser entrance such as from the side or back of the building. This then gives way to the manifestation of negativity towards Islam in the eyes of the younger generation of Muslims. Non-Muslims will equally have a bad impression of what Islam really is. The idea of Islam marginalizing and isolating women will become prevalent catalyzed from poor spatial planning and consideration. Hence, why the inclusion of women within the mosque boundary is essential and encouraged and should be modeled and inspired after how women are treated during the Prophets (pbuh) time. In addressing this pressing issue, the research will focus on the participation and activities held in a women's space for prayer during the Prophet's (pbuh) time.

3. Importance of women in Islam

Prof. Abdur Rahman I. Doi said in Islam there is absolutely no difference between men and women as far as their relationship to Allah is concerned, as both are promised the same reward for good conduct and the same punishment for evil conduct. The Quran says:

"*And for women are rights over men similar to those of men over women.*" [Noble Quran 2:228] He further stated in his book that the Quran admonishes those men who oppress or ill-treat women:

"*O you who believe! You are forbidden to inherit women against their will. Nor should you treat them with harshness, that you may take away part of the dowry you have given them except when they have become guilty of open lewdness. On the contrary live with them on a footing of kindness and equity. If you take a dislike to them, it may be that you dislike something and Allah will bring about through it a great deal of good.*" [Noble Quran 4:19]

The Prophet (pbuh) stated that women are 'twin halves' of men. The Quran emphasizes the essential unity of men and women in a most beautiful simile:

"*They (your wives) are your garment and you are a garment for them.*" [Noble Quran 2:187] On other occasions, the Prophet (pbuh) said:

"*The best property a man can have is a remembering tongue (about Allah), a grateful heart and a believing wife who helps him in his faith.*"

And again:

"*The world, the whole of it, is a commodity and the best of the commodities of the world is a virtuous wife.*"

Women were considered to have as much value as animals before the beginning of Islam and the Prophet (pbuh) wanted to stop all inflicted cruelties towards women. He told the Muslims: "*Fear Allah in respect of women.*"

And:

"*The best of you are they who behave best to their wives.*" And:

"*A Muslim must not hate his wife, and if he be displeased with one bad quality in her, let him be pleased with one that is good.*"

And:

"The more civil and kind a Muslim is to his wife, the more perfect in faith he is."

In his book, he further describes that The Prophet (pbuh) was most emphatic in enjoining upon Muslims to be kind to their womenfolk when he delivered his famous *khutbah* on Mount of Mercy at Arafat in the presence of one hundred and twenty-four thousand of his *Sahabat,* who gathered there for the *Hajj al-Wada* (Farewell Pilgrimage). In it, he ordered those present, and through them all those Muslims who were to come later, to be respectful and kind towards women. He said:

"Fear Allah regarding women. Verily you have married them with the trust of Allah, and made their bodies lawful with the word of Allah. You have got (rights) over them, and they have got (rights) over you in respect of their food and clothing according to your means."

Dr. S. Ansiya Begum, Beena M.N (2010), also elaborately stated the importance of women in her research. She stated that in Islam, there is no sex based distinction in spiritual matters. The paradise and its blessings are meant the same for both men and women. The five tenants of Islam (belief, prayer, fasting, poor due, and the pilgrimage) do not differentiate men and women. To explain further, Islam encourages and asks that education of women be at par with men. The Quran condemned the pre-Islamic practice of female infanticide, also women have right to property and justice; she has the right to maintenance, widowed or divorced, the women are eligible for remarriage. As clearly as this is stated within the Quran, it shows the importance and status of a female in the Islamic society.

According to Naveed Ahmad Lone, Hurmat Sumaiya Bint Bashir, Yousfi Karima (2015), the status of women in Islam is clear and unambiguous. The aptitude of Islam as delineated in the basic manuscripts, i.e. The Quran and Hadith bear witness to the fact that women is as important as men and that she is not inferior to him. The Quran and Hadith have discussed that women should be given her natural status and how Islam has empowered women from all aspects of life.

4. The Current Situation for Women's Access and Participation in Mosques

Despite the fact that Islam is a communal religion — a brother and sisterhood — making women feel unwelcome at the mosque is endemic in the Muslim world. Particularly, how many mosque prayer spaces are not child friendly, and when they are, the onus is placed on mothers to take care of the children while the men are free to worship. Naturally, when children are excluded, it means that women are excluded from religious communal worship. Islam does not forbid women from going to the mosque. In fact, it was encouraged by the Prophet (pbuh) when he said:

"Do not prevent the female servants of Allah from going to the mosque…" (Muslim, Abu Dawud).

The mosque is the center of the community of access to Islamic education and is key to fulfilling certain religious rites. The traditions of the Prophet (pbuh) tell us that any place on Earth is suitable for prayer, any place except anything blatantly inappropriate. It is also in these traditions however, and particularly in the subsequent centuries of predominantly male interpretations of these traditions, that we find support for a whole slew of places for women to pray; behind men, above men in balconies, beneath men in mosque basements, or just not being allowed in the mosque. (Sana Fatima, 2016)

Women who are traveling, working, unable to be at home or another mosque to pray, the scope for them to find a suitable space in mosque for offering prayer is not easy. The majority of mosques all over the world do not have well designed prayer space for women. The prayer spaces are in either the basement or a left over spaces, which have insufficient area, poorly ventilated or derelict. There is a void in almost all the community and multitude of viewpoints, ideas, and energy have been eliminated. More than 50% of the local community is invisible and excluded. It is 50% because it is usually the case that when a mosque excludes women, it automatically excludes children as well.

It was not the face of the *Da'wa* (invitation) of the *Sahaba* (companions) and the *Sunnah* (way of the Prophet (pbuh)) of the Prophet (pbuh) to exclude women, not from the mosque and not from the community as a whole. Much is made of the *Hadith* (narration of the life of the Prophet (pbuh)) in which the Prophet (pbuh) told a woman that prayer in her home is better than prayer in the mosque. Religious teacher Ustaz Abdullah Adhami taught this *Hadith* in line with common sense and a traditional point of view where he discusses misinterpretations made of this *Hadith* to justify banning women from mosques. While the Prophet (pbuh) told the woman that the prayer in her home is better for her, he did not forbid her from coming to the mosque at all. In fact, we know that the contrary is true - that he forbade men from preventing women to go to the Masjid, as seen in the numerous *Hadith*. (Abdul Malik Mujahid, 2016)

In Muslim countries, mosques that designate areas for women's prayer rarely allow women to line up directly behind men in the same hall, as was the practice in the Prophet's (pbuh) Mosque during his lifetime. In many Arab countries, it is common to separate men's prayer place from women's, especially in small mosques. The distance between the two prayer areas varies from one mosque to another. The majority of mosques in Muslim countries contain special halls or small rooms for women in the basement, in the ground floor, in a closed balcony, or in a small building attached to the mosque. Speakers are used to communicate the Imam's voice in prayer. The problem in confining women to such halls is manifold. First, these halls are much smaller than the main prayer hall, usually overcrowded, especially during the Friday Prayers and other occasions the men's prayer areas and mosque halls are far from full. Women outnumber men in public occasions, especially in countries with Muslim minorities. Second, these halls are not properly equipped and furnished as men's halls are, in terms of carpets, lights and sound devices. Hence, women feel less welcomed and less privileged than men in mosques do. Third, the places assigned for children in these mosques are attached only to women halls, which cause much distraction for them. Fourth, women are not allowed to enter mosques through their main gates; rather, they would alternatively use narrow side entrances or secondary entrances through the back of the mosque. Due to the issues discussed, the design of the mosques leads to an impression from non-Muslims and the younger generation of Muslims that Islam marginalizes and isolates women within the religion as evident as the separation of women within the mosque itself. Noticeably the incommodious, poorly furnished and noisy prayer halls assigned to women reinforces the sentiments that Islam does not approve of a woman's presence in a place of prayer and that a man in Islam do not share the burden of caring for the child and is the religion that upholds patriarchy as is claimed by some non- Muslims. (Jasser Auda, 2016)

According to Hind Makki (2012), in most Western Muslim communities today, women are not prevented from offering their prayers in mosques. They are even encouraged to attend, especially during Friday prayers and the nightly *Taraweeh* prayers that Sunni Muslims offer during the holy month of *Ramadan*. However, it often seems that the Prophetic tradition invoked in the story above is the only safety net women have. Too many mosques have inadequate spaces for female congregants; others have the space, but keep the doors locked. Still others have adequate spaces for women to pray, but assume the women's prayer hall as the space to be kept aside for late- arriving men to use, as according to the Islamic belief, God forbade the men to pray alongside with the women and children.

Yet, poorly designed spaces for women in Mosque space are continuously built. In every situation, believers use their conscience by giving importance to primarily, a guest's provision to pray rather than being judgmental over gender. For example, if the men's prayer hall becomes full and space remains in the women's, the remaining empty space should be utilized to ensure the arriving congregation a space of prayer, rather than be turned away from the Mosque and prevented from praying altogether - ("...who bar people from the places of worship..." {2:114}) - and understanding that prayer and serving Allah should be prioritized, rather than being prejudiced over gender. As the Quran, further states;

"It is not suitable that believers being friends and protectors of one another (9:71) ignore the need of the other but rather should be courteous in ensuring they have the place to perform prayer."

"Have you considered the one who prevents our servant from praying? Have you considered whether he is truly guided or not? Or commands (people) to true righteousness?" (Qur'an 96:9-12)

Many Muslim women are accustomed to finding themselves in a situation where it is prayer time, but the Mosque in front of them does not welcome female visitors. They are consistently in uncertainty and left wondering whether they should inquire, or skip prayer altogether. It can even be very daunting approaching a male worshipper for assistance, as they can be inhospitable. (Hasan. M. and S. Sayeed, 2014)

Islamic Social Services Associations and Women in Islam, Inc. have also discussed about the results of the Council of American Islamic Relations (CAIR) 2001 Report from the Mosque Study Project (MSP) and their own qualitative interviews with many women on the status of women's prayer space in mosque. The MSP shows a general picture of practices across mosques. The interviews capture women's experiences of mosque practices.

5. Mosque Study Project Results

The MSP, co-sponsored by the Council of American Islamic Relations (CAIR), Islamic Society of North America (ISNA), Ministry of Imam W. Deen Mohammed, and the Islamic Circle of North America (ICNA), surveyed 416 randomly sampled mosques in the United States. Results showed that:

The majority of those who regularly participate in mosques are men. The MSP showed that on average, across most mosques, 75% of regular participants are male. Although 54% of mosques reported regular activities for women, a sizeable majority reported only occasional activity (27%) and 19% said they did not offer any programs for women. While 50% of mosques report that women have served at one time or another on their governing or executive boards, a sizeable proportion of mosques still prevent women from serving on their executive boards (31%). Nineteen percent said they allowed women to serve, but did not have any women actually serving on a board for the past five years. The practice of women praying behind a curtain or in another room has increased. In 1994, 52% of mosques reported that women make prayers behind a partition or in another room, but that practice was adopted by 66% of mosques in 2000. Most regular participants are 36 years or older (53%).

6. Women's Experiences and Perceptions of Mosque Practices

In interviews across Canadian cities (conducted in 2001 by ISSA) many women voiced broad concerns, as well as problems of access, participation, and the approaches of mosque leadership. Women In Islam, Inc. also receives regular reports of gender discrimination in mosques and other community institutions. Many women believe that cultural biases and mindsets, rather than principles of Islam, are at the root of how they are treated at the mosque and at community centers. Of course, women recognize that some mosque leaders may be well meaning, but are unable to harness the financial resources that are needed to build open spaces, maintain structures, or develop quality programs. However, they also thought that all communities can benefit from self-reflection and self-evaluation regarding mosque practices that affect women in particular and the community in general.

7. Experiences and Perceptions Related to Access

Since education in Islam is not only a right, but also a duty of all Muslims, women, too, are eager for religious knowledge, and enjoy learning from imams, scholars and other speakers. However, some felt their education was impeded when they were unable to attend the mosque, hear or see the speakers, or they were not given an opportunity to ask questions. Some reported they were discouraged from and even rebuked for entering the mosque library. Qualified women scholars and other professional and activist women are not invited to speak; women feel that they, their children, and indeed their communities, would benefit tremendously from contact with such role models who are maintaining their *deen* as well as advancing in their fields.

8. Experiences and Perceptions about Mosque Governance and Mosque Participation

According to the Islamic Social Services Associations and Women in Islam, Inc., although more women participate in mosque activities, some also simultaneously felt that the dignity and honor given to them by Islam is not reflected in the treatment they encounter at their community mosque. For example, women who have tried to approach mosque leadership have sometimes found them dismissive of women's ideas and concerns. When women are invited to take managerial roles in mosque programs, they are often limited to serving as cooks and cleaners; while women enjoy taking care of their communities in these ways, they felt leadership could do more to include women in mosque consultation processes or meetings. They believed that it was cultural gender biases that limited their religious right to have a role in decision-making. Mothers with children said that they would feel more encouraged to attend the mosque if there were designated play areas for children, provided there are community resources to build these spaces. The low attendance of women adversely affects unmarried, divorced, and widowed women. More specifically, they feel a greater sense of isolation at the mosque and find it harder to maintain contact and integrate with the wider community. These women have a right to special community support. Women felt that standards of modest dress are not applied consistently for men and boys as they are for women. Lastly, within the North American context, women perceive that the *khutba* (Friday sermon) must become more sensitive to the language and culture of North America and are not balanced in their content. Gender issues, when addressed in the *khutba*, must be discussed in ways that highlight the differences between culture and religion and recognize the diversity of Muslim women's experiences.

9. Suggestions for Overcoming the Problem and to Encourage Women to Participate in Mosque and Other Social Activities

A mosque that is open to women's access and participation is a mosque that welcomes present and future generations of Muslims. As noted in the Quran,

"And as for the believers, both men and women they are helping friends (Awlia) of each other, they enjoin the doing of good and just and forbid what is evil and unjust" (9: 71).

According to Yasmine Hafiz (2013), if the mosque space is truly very small, there are very easy ways to make it available to women who need to pray there, while opening up the entire space for the men when no women are present. Many home improvement and home decorating stores sell decorative screens (like the rice paper ones seen in Japan), for a relatively low price. It folds up and are easy to store when not in use. Office supply stores sell cubicle walls with wheels. It also folds up for easy storage. If the mosque does not have enough budget for these items, then taking up a special collection and install an extended curtain rod across the intended space for women and put up floor length curtains. Tables or chairs can be used to mark the space reserved for women as mosques have been doing this for hundreds of years and that is enough to designate a space behind the men as women's space, without barriers or walls. However, it should also be kept in mind that some women and men might not be comfortable with this arrangement, since there are women who may need to breastfeed an infant or adjust their coverings in the course of a *Juma'ah khutba*.

If there is larger amount of space in the mosque, then the space should be reserved for the women and should be adequate. The floor should be clean. It must be heated in the winter and has air in the summer. The roof should not leak when it rains. And there should be copies of the Quran are on hand of women to read. The women's bathroom must have hooks for their hijabs (when they are making *wudhu*), paper towels for them to dry with, slippers to wear, and soap to wash with. The bathroom must be clean. Where there are Muslim women, there are bound to be Muslim children and the smallest of those children will need to have their diapers changed. Ensuring that women are included in the mosque means ensuring that the entire community has access to the teachings of Islam. It shows non-Muslims that Islam does not stand for the exclusion of women and children, that Islam is not a "man's religion." It shows non-Muslims that a woman can be modest, can be religious and still participate in community life. It shows the next generation of Muslims that cultural ideas about excluding women and keeping them in the home are not from Islam and following the teachings and example of the Prophet (pbuh). The understanding of pre- Islamic cultural idea of what a 'women's place' is defined as is slowly unraveling with the new dynamics of women within the Islamic society. The misconceptions and misunderstandings of the *Deen* (religion) is also changing with further studies of what Islam really stands for in terms of the role of women. The only way that we can be sure that the next generation understands Islam as it was truly taught by the Prophet (Peace be upon Him) is to be sure that women and children are fully included in the mosque. (Abdul Malik Mujahid, 2016)

According to the Islamic Social Services Associations and Women In Islam Inc., below are some suggestions for how to modify physical access, participation, and mosque governance. Some of these changes may be harder to achieve than others, depending on community's resources, but nonetheless important. In addition, each mosque should conduct a systematic inventory of its current practices in terms of women's access to mosque space, program planning and participation, and mosque governance and management.

10. Conclusion

Due to the above limitations in current mosque practices, many women go to the mosque with some trepidation, and some go only occasionally or do not attend at all. One of the unfortunate consequences is that women and their children are losing the knowledge about *adab al mosque* or rules of attending the mosque and congregational prayers. In addition, an entire generation of

IOP Conf. Series: Materials Science and Engineering **401** (2018) 012032 doi:10.1088/1757-899X/401/1/012032

women and children are moving away from the mosque and finding other avenues for their intellect and training. Mosque leaders who do not attend to women's inquiry and give input are also missing the benefit of women's perspective and ideas for community improvement. A status quo that excludes women also makes them and their children less able to practice their community responsibilities to promote good and forbid evil.

These experiences are not in accordance with Islam. The practice of our beloved Prophet (pbuh) was to encourage active learning and questions by all members of the community. In addition, a mosque and other facilities such as libraries, belong to the community, and therefore must be accessible to all community members, including women.

The alienation of women from the mosque should be addressed at the community level. Each mosque must gradually but in a determined fashion, modify its architecture, governance, and programs to be inclusive of women and children. The leadership at each mosque must be proactive in initiating and supporting these changes.

Reference

[1] Chowdhury, T. (n.d.). Segregation of Women in Islamic Cultures and Its Reflection in Housing: A Study of Spaces for Women in a Bangladesh Village. Doi, A. R. (2015). Islam and Women.

[2] ESKANDARI, M. (n.d.). Women's Places and Spaces in Contemporary.Islamic Social Services Associations and Women In Islam, I. (n.d.). Women Friendly Mosques and Community Centers: Working Together to Reclaim Our Heritage.

[3] Karim, B. R. (2015). Effective Utilization of Our Mosques. Mahboubeh S.

[4] Habani Jafroudi1, M. K. (N.D.). Analysing Persistence and Formation of Social Interactions in Religious Space Of Mosques. *Indian Journal of Fundamental and Applied Life Sciences ISSN: 2231– 6345 (Online)* . Yahya, H. (2015). Women in the Quran. *The importance of Women* .

[5] Baharudin, N. '., 2014. Communal Mosques: Design functionality towards the development of sustainability for community. *AMER International Conference on Quality of Life.*

[6] Campo, J. E., n.d. *Encyclopedia of Islam.* s.l.:s.n. Creswell, K. A. C., 1989. *A Short Account of Early muslim Architecture.* london: Scholar Press.

[7] Farahati, A., 2011. The Mosque as the First Political-Ideological Base in the Islamic Society. *Journal of Politics and Law,* 4(1).

IOP Conf. Series: Materials Science and Engineering **401** (2018) 012033 doi:10.1088/1757-899X/401/1/012033

Review of Recycles Concrete Aggregate and Polyethylene Terephthalate in the Manufacturing of Brick

P N Mazenan[1], F S Khalid[1*], N B Azmi[1], S S Ayop[1] A H Abdul Ghani[1] and J M Irwan[1]

[1] Department of Structural and Material, Faculty of Civil and Environmental Engineering, Universiti Tun Hussein Onn Malaysia, Batu Pahat, Johor.

Corresponding e-mail: faisalsh@uthm.edu.my

Abstract. Recycled Concrete Aggregate (RCA) and Polyethylene Terephthalate (PET) are one of the construction waste can be recycled. Both of them can be the aggregate to replace the natural aggregate in concrete since we know the physical properties of both materials are hard and strong. Demand for sand in the concrete production has increased which led to become problems in the industry. As a result, the price of the sand increased, which in turn affected the pricing of the cement and sand bricks. One of the ways to reduce this problem is to utilize the RCA and PET waste bottle in the production of construction materials such as brick. Replacing both wastes as sand aggregate in bricks is one of practices and imaginative solutions in order to minimize environmental impacts. In short, using 50% replacement of RCA as replacement of fine aggregate show the acceptable in strength of brick. Furthermore, the decision of using PET will make it worthwhile in increasing the workability of economical brick.

1. Introduction

Nowadays, there are various type of aggregate which has been produced that consist of materials, industrial waste, crushed glass, construction waste such as concrete, crushed brick and more [1]. Classification of aggregate is generally based on their geological origin, size, shape and unit weight [2]. Recycled aggregate consists of original aggregate and original mortar. Hence, it is using the aggregate refining method in production for the purpose of reduce the quality of using natural aggregate such as gravel or sand [2]. RCA obtained from crushed concrete that containing of 65-80 % of natural coarse aggregate and fine aggregate and 20-35 % of old cement [3]. According to [1]Sheikh Khalid *et al.* [4] study, when RCA is crushed from masonry, the content of the fine particles is much more than that form concrete. This leads to decrease in strength and workability of concrete containing recycled fine aggregate.

Recycled concrete aggregate produced by two stage which is start by crushing of demolished concrete, screening and removal of contaminants such as reinforcement, paper, wood, plastic and gypsum [5]. However, one of the problems exists when using the RCA for manufacturing of new concrete mixtures is the possibility of contaminants in original construction and demolition (CDW) debris passing into new concrete mixtures. The contaminants may be organic material, glass, lightweight bricks, chemical admixtures, clay and other metals [5]. The previous researcher found that 25% and 50% replacement of recycled aggregate in concrete brick and paving block showed the increase percentage of replacement was reduced compressive strength in production of samples [6].

Content from this work may be used under the terms of the Creative Commons Attribution 3.0 licence. Any further distribution of this work must maintain attribution to the author(s) and the title of the work, journal citation and DOI.
Published under licence by IOP Publishing Ltd

Previous published paper explain which tells the cost of transportation of natural resources, energy and reduces the impact of waste material on industry environment [7]. However, the researcher claims that it would be inexpensive by using coarse and fine recycled aggregate as new construction materials [8].

2. RCA in Brick Production

The result of density sand cement brick shows that, the decreasing value of density at RCA 75% by 4.9% for 7 days and 5.4% for 28 days which are lower compared to control brick. However, 55% of recycled concrete aggregate is added showing that, the value of compressive strength decreases due to smaller porosity and is less permeable compared to control brick. High percentage of recycled concrete aggregate in sand cement brick was more permeable than the natural sand [9].

Bravo *et al.* [10] studied the durability performance of recycled concrete aggregate in brick. They used different mixes with 10%, 25%, 50% and 100% for volume of recycled concrete aggregate replacement of coarse natural aggregate (from five sources). Meanwhile, the same percentage had been used for volume of fine natural aggregate replaced by fine recycled aggregate. From this study, the water absorption was analysed that it can be concluded that the replacement of NA with RA caused an increase of the water absorption in samples. Dina M. & Mohamed M. [11] stated that concrete containing coarse crushed bricks had a relatively lower strength at early ages than normal aggregate concrete. Compressive strength, water absorption and unit weight of solid cement bricks containing crushed brick aggregates were determined and compared with the limits of the national and international standards for load bearing and non-load bearing units. The results obtained that specific gravity of brick aggregates had lower specific gravity compare to natural aggregates. However, the water absorption showed the higher value was 40% than natural aggregates due to porosity of the clay bricks.

Based on Ismail and Yaacob [12], this study aims the possibility of recycled fine aggregate produced from waste as a substitution for natural aggregate in brick. The percentage of recycled aggregate used were 0%, 25%, 50%, 75% and 100%. The replacement of recycled fine aggregates at level 50% and 75% showed a good value on compressive strength in replacement of natural sand. Hence, in flexural strength test, the result of 50% replacement recycled aggregate showed slight decrease which due to porously of recycled aggregate in brick. However, water absorption and shrinkage results of the brick shows a good result. Poon *et al.* [13] studies aims to using RCA in sand cement bricks. A total series of mixtures was remove repared according to the replacement of RCA in natural aggregate of 0%, 25%, 50%, 75% and 100%. The test that involved in their studies is compressive strength, drying shrinkage and transverse strength test. The results obtained, the compressive strength for brick shows that the percentages of 50% have a greater value rather than the control brick. However, the result shows a reduction when the percentage of RCA is 75% and 100%. Transverse strength test presents that all percentage of RCA shows a high value compare to control and its pattern similar with drying shrinkage test. They indicated that the although percentage of 25% and 50% of RCA shows an impact on compressive strength, transverse strength of the specimen increased as percentage of RCA increased. Table 1 shows the overall summary of previous study on RCA.

Table 1. Summary of previous work on RCA.

Author	Percentage of replacement	Description
Sheikh Khalid, F et al., 2017[9]	55%, 65%, and 75%.	• Replacement by weight of recycled concrete aggregate were 55%, 65%, and 75% • Compressive strength, water absorption and density test • The density of sand bricks containing recycled fine aggregate decreases when the percentages of recycled fine aggregate increases. • The value of compressive strength decreases but it shows less permeable compared to control brick when 55% of RCA is added

Bravo et al., 2015 [10]	10%, 25%, 50% and 100%	• The replacement of 10%, 25%, 50% and 100% of the overall volume of coarse and fine natural aggregate was replaced • The water absorption can be concluded that the replacement of NA with RA caused an increase of the water absorption in samples.
Dina et al., 2012 [11]	0%, and 25%	• The substitution ratios of natural aggregates with RCA were 0% and 25% by volume. • Four series of mixtures with cement content 100, 150, 200 and 300 kg/m³ were prepared • The water absorption increased and the unit weight of solid cement bricks decreased • Bricks that containing crushed aggregates present the decreasing value of compressive strength due to increasing the size and content of crushed bricks.
Ismail, 2010[12]	25%, 50%, 75%, and 100%	• The design proportion was used 90% sand, 5% cement and 5% laterite soil of the cement and sand bricks • The replacement were 25%, 50%, 75%, and 100% by weight of natural sand aggregates. • The higher replacement of recycled fine aggregate, the lower the density of cement and sand bricks • 50% and 75% replacement showed a good value on compressive strength in replacement of natural sand. • Water absorption and shrinkage results of the brick shows a good result.
Poon et al., 2002 [13]	25%, 50%, and 100%	• Natural aggregate was replaced at level from 25% to 100% by weight. • Compressive strength showed a little effect at level 25% and 50% when coarse and fine natural aggregates was replaced to the bricks and blocks sample. • The increase levels of replacement reduced the compressive strength of samples.

3. PET Waste in Brick Production

PET stands for Polyethylene Terephthalate is usual plastics which is used as a raw material for making products like blown bottle which is used to produce soft drinks, food containers and other consumer goods. [14]. These non-biodegradable waste materials along with population growth have caused a global environmental crisis all around the world. In order to protect the environment, the new technologies to recycle and convert waste materials have to be developed. This leads to waste materials into reusable materials that important to sustainable development of society. The present application of recycled PET waste in the construction industry includes its use as resin for polymer concrete [15] and as fiber for reinforced concrete [16]. PET waste can be substitute as aggregate to reduce the waste in order to bring the economical way in construction industry. Moreover, the using PET waste in concrete has a few advantages such as reduce the usage of natural resources, decrease the waste consumption and protect the environment economizing energy and pollution [10,17].

Vadivel *et al.* [18], used the volume of PET of 0%, 5%, 10%, 15%, 20% and 25% as sand replacement and the compressive strength, flexural strength and split tensile strength test. The compressive strength of PET gradually increases until volume of 15% but it starts to decrease in 20% and 25%. The replacement of 15% PET shows high value compare to the normal concrete. Split tensile strength of PET performed very well compared to conventional concrete. The volume of 25% PET shows high split tensile strength which is 24.87% than conventional. The flexural strength of PET shows a good result rather than the normal concrete where the 25% replacement of PET provides the higher value. Silva *et al.* [19] study the effect of curing conditions on the durability of concrete mixes with selected plastic waste aggregates. Tests for chloride penetration, carbonation, water absorption and compressive strength were carried out to determine the durability of each sample. Low bonding strength between PET and aggregates make the reduction of strength concrete due to the PA's impermeable nature. The coarser the particle size of the PA lead to reduction in the level of packing concrete and the aggregate leading to poorer bonding of concrete when water absorbed by the PA.

Ge *et al.* [20] study the effects of aggregate gradation, sand-to-Polyethylene Terephthalate (PET) ratio and curing conditions on physical and mechanical properties of recycled PET mortar. Results show that sand-to-PET ratio increased, the density and water absorption of the recycled PET mortar also increased. Similar observations conducted by Ghernouti *et al.* [21], where the results showed that the use of plastics bag wastes (PBW) containing 10 and 20% of waste enabled to reduce by 18-23% the compressive strength of mortars, which remains close to the reference mortar (without PBW waste). The author used (10, 20, 30 and 40 %) of PBW as sand replacement aggregate.

Shahidan *et al.* [22] focus this study on physical and mechanical behavior of recycled PET fiber reinforced mortar. This study claims that, PET bottle was adding about 0%, 0.5%, 1.0% and 1.5%, in dry mix mortar remarkably better to improve strength of concrete. The best replacement of 1.0 % PET fiber is the best maximum volume in production of sample. Some previous studies conducted by Akcaozog˘lu *et al.* [23] used 0.45 water binder ratio and 0.50% PET in the mixtures. This study aimed the physical and mechanical properties of mortars containing PET as replacement of aggregates. The test concluded such as the compressive strength, water absorption and carbonation tests. The mixtures were prepared by using only waste PET granules and half of the mortar mixture sand the rest mortar mixtures were prepared with sand and waste PET granules together. The results showed that the mixtures containing sand and PET were higher in compressive strength compared to the mixtures containing PET without sand.

Choi *et al.* [24] studied the performance of properties of waste PET lightweight aggregates (WPLA) in mortar and concrete. In this study, the author used 0%, 25%, 50%, 75% and 100% volume of WPLA and the test involved was density, compressive strength and water absorption tests were conducted. The mortar and concrete mixtures were used with water cement ratio 0.6. The compressive strength with 0% WPLA 100% showed the value 44.9 Mpa at strength 28 days. However, the compressive strength at 28 days reduced as the percentage of WPLA replacement was increased. The replacement of 100% WPLA showed the compressive strength at 28 days slightly reduce 42% of mortar samples. Other researchers studied the behavior of concretes reinforced with PET bottles waste fibers. It was concluded that adding small quantity of PET bottle fibers have a noticeable effect on post-cracking behavior of concretes. Additionally, PET fibers enhance the toughness and improve the concrete plasticity [25-26]. Moreover, it can be concluded that the increase percentage of WPLA, the more the compressive strength of the mortar decrease. Summary of previous work on waste PET are shown in Table 2.

Table 2. Summary of previous work on waste PET.

Author	Max. size of PET (mm)	W/c ratio (%)	Volume of PET (%)	Test Involved	Description
Vadivel et al., 2016 [18]	<5	0, 5, 10, 15, 20, 25	0, 5, 10, 15, 20 and 25	compressive strength, split tensile strength and flexural strength test.	The higher replacement of PET decreases the strength of concrete compare to normal concrete

Silva et al., 2015 [19]	<4	0.52 to 0.61	0, 7.5 and 15	Compressive strength, water absorption by immersion	The compressive strength of each concrete mix decreased as the plastic aggregate (PA) content increased
Ge et al., 2012 [20]	<5	0.5	10, 20, 30 and 40	Compressive Strength	Compressive strength decreases by 18–23 % of mortars containing 10 and 20 % of waste PET respectively
Shahidan et al., 2018 [22]	5	0.45	0, 0.5, 1.0, 1.5 and 2	Measured the compressive and splitting tensile strength of concrete	The best replacement of 1.0% PET fiber is the best maximum volume in production of concrete sample
Akcaozog˘lu et al., 2010 [23]	<4	0.45	0.50	Compressive strength, water absorption and carbonation	The samples containing sand and PET were higher in compressive strength and flexural strength compared to the samples containing PET without natural sand.
Choi et al., 2009 [24]	0.15	0.6	0, 25, 50, 75 and 100	Density, compressive strength and water absorption	The higher percentage of WPLA, the more compressive strength of the mortar decrease

4. Conclusion
From the researchers discussion above, it is clearly shown that Recycled Concrete Aggregate (RCA) and Polyethylene Terephthalate (PET) are suitable to use in the construction industry and more economic in the making of brick. RCA and PET are found to be suitable as a partial replacement of fine aggregate in brick production. Most of the researchers suggested up to 50% replacement of RCA as a fine aggregate replacement in the brick production. Replacement more than 50% leads to the decreases in the strength of brick. For PET waste, the increasing replacement percentage of PET will decrease the mechanical properties of brick respectively.

5. References
[1] Kubissa W, Jaskulski R, Koper A and Szpetulski J 2015 Properties of Concretes with Natural Aggregate Improved by RCA Addition *Procedia Engineering* **108** pp 30–38

[2] Khalid F S, Azmi N B, Mohd Sumandi K A S and Mazenan P N 2017 Mechanical properties of concrete containing recycled concrete aggregate (RCA) and ceramic waste as coarse aggregate replacement *AIP Conference Proceedings* **1891** 020079

[3] Kien T T, Thanh L T and Lu P V 2013 Recycling construction demolition waste in the world and in Vietnam *In The International Conference on Sustainable Built Environment for Now and the Future* Hanoi **26** p 27

[4] Sheikh Khalid F, Herman H S, Bazilah Azmi N 2017 Properties of Sugarcane Fiber on the

Strength of the Normal and Lightweight Concrete *MATEC Web of Conferences* **103** 01021

[5] Malešev M, Radonjanin V and Marinković S 2010 Recycled concrete as aggregate for structural concrete production *Sustainability* **2(5)** pp 1204-1225

[6] Ismail S and Yaacob Z 2010 Properties of bricks produced with recycled fine aggregate *International Journal of Civil, Environmental, Structural, Construction and Architectural Engineering* **4(7)** pp 190-194

[7] Khalid F S, Irwan J M, Wan Ibrahim M H, Othman N and Shahidan S 2018 Splitting tensile and pullout behavior of synthetic wastes as fiber-reinforced concrete *Construction and Building Materials* **171** pp 54-64

[8] Khatib J M 2005 Properties of concrete incorporating fine recycled aggregate *Cement and Concrete Research* **35(4)** pp 763-769

[9] Sheikh Khalid F, Shah Herman H, Azmi N B and Juki M I 2017 Sand Cement Brick Containing Recycled Concrete Aggregate as Fine-Aggregate Replacement *MATEC Web of Conferences* **103(5)** 01016

[10] Bravo M, De Brito J, Pontes J, and Evangelista L 2015 Durability performance of concrete with recycled aggregates from construction and demolition waste plants *Construction and Building Materials* **77** pp 357-369

[11] Dina M S and Mohamed M E 2012 Development of High-Performance Green Concrete using Demolition and Industrial Wastes for Sustainable Construction *Journal of American Science* **8(4)** pp 12-131

[12] Ismail S and Yaacob Z 2010 Properties of bricks produced with recycled fine aggregate *International Journal of Civil, Environmental, Structural, Construction and Architectural Engineering* **4(7)** pp 190-194

[13] Poon C S, Kou S C and Lam L 2002 Use of recycled aggregates in molded concrete bricks and blocks *Construction and Building Materials* **16(5)** 281-289

[14] Ali N, Din N, Khalid F S, Shahidan S, Abdullah S R, Samad A A A and Mohamad N 2017 Compressive strength and initial water absorption rate for cement brick containing high-density polyethylene (HDPE) as a substitutional material for sand *IOP Conf. Series: Materials Science and Engineering* **271** 012083

[15] Khalid F S, Azmi N B, Mazenan P N, Shahidan S, Othman N H and Anak Guntor N A 2018 Self-Consolidating Concretes Containing Waste PET Bottles as Sand Replacement *AIP Conference Proceedings* **1930** 020034

[16] Irwan J M, Asyraf R M, Othman N, Koh H B, Aeslina A K, Annas M M K, Faisal S K 2014 Deflection behaviour of irregular-shaped Polyethylene Terephthalate fibre reinforced concrete beam *Advanced Materials Research* **911** pp 438-442

[17] Faisal Sheikh Khalid, Mohd Irwan Juki, Norzila Othman and Mohd Haziman Wan Ibrahim 2017 Pull-out strength of polyethylene terephthalate bottle fibre in concrete Matrix *Malaysian Construction Research Journal (MCRJ)* **21(1)** pp 75-85

[18] Vadivel S T and Doddurani M 2013 An Experimental Study on Mechanical properties of Waste Plastic Fiber Reinforced Concrete *International Journal of Emerging Trends in Engineering and Development* **3(2)** pp 395-401

[19] Silva R V, Neves R, De Brito J and Dhir R K 2015 Cement & Concrete Composites Carbonation behaviour of recycled aggregate concrete *Cement and Concrete Composites* **62** pp 22–32

[20] Ge Z, Yue H and Sun R 2015 Properties of mortar produced with recycled claybrick aggregate and PET *Construction and Building Materials* **93** 851856

[21] Ghernouti Y and Rabehi B 2012 Strength and Durability of Mortar Made with Plastics Bag Waste (MPBW) *International Journal of Concrete Structures and Materials* **6(3)** pp 145–153

[22] Shahidan S, Ranle N A, Zuki S S M, Khalid F S, Ridzuan A R M and Nazri F M 2018 Concrete incorporated with optimum percentages of recycled polyethylene terephthalate (PET) bottle fiber *International Journal of Integrated Engineering* **10(1)** pp 1-8

[23] Akcaozoglu S, Atis C D and Akcaozoglu K 2010 An investigation on the use of shredded pet waste bottles as aggregate in lightweight concrete *Waste Management* **30** pp 285–90

[24] Choi Y W, Moon D J, Kim Y J and Lachemi M 2009 Characteristics of mortar and concrete

containing fine aggregate manufactured from recycled waste polyethylene terephthalate bottles *Construction and Building Materials* **23(8)** pp 2829–2835

[25] Khalid F S, Juki M I, Othman N and Wan Ibrahim M H 2016 Flexural toughness of ring-shaped waste bottle fiber concrete *MATEC Web of Conferences* **47** 01002

[26] Irwan J M, Annas M M K, Aeslina A K, Asyraf R M and Faisal S K 2013 Development of concrete mix design nomograph containing polyethylene terephtalate (PET) as fine aggregate *Advanced Materials Research* **701** pp 12-16

Acknowledgments
This work was financially supported by University Tun Hussein Onn Malaysia (UTHM) and Minister of Education Malaysia (KPM) through Dana Pembudayaan penyelidikan (RAGS) Vot. No. R-058

Institute of Physics
Dirac House, Temple Back
Bristol BS1 6BE UK

ISSN: 1757-8981
ISBN 978-1-5108-7255-4

CURRAN ASSOCIATES INC.
proceedings.com